# From Variability Tolerance to Approximate Computing in Parallel Integrated Architectures and Accelerators

Abbas Rahimi · Luca Benini
Rajesh K. Gupta

# From Variability Tolerance to Approximate Computing in Parallel Integrated Architectures and Accelerators

Springer

Abbas Rahimi
Department of Electrical Engineering and Computer Sciences
University of California Berkeley
Berkeley, CA
USA

Rajesh K. Gupta
Department of Computer Science and Engineering
University of California, San Diego
La Jolla, CA
USA

Luca Benini
Integrated Systems Laboratory
ETH Zurich
Zürich
Switzerland

ISBN 978-3-319-53767-2 ISBN 978-3-319-53768-9 (eBook)
DOI 10.1007/978-3-319-53768-9

Library of Congress Control Number: 2017932004



Printed on acid-free paper

This Springer imprint is published by Springer Nature
The registered company is Springer International Publishing AG
The registered company address is: Gewerbestrasse 11, 6330 Cham, Switzerland

*To my wife with everlasting love and gratitude*

–Abbas Rahimi

# Foreword

There is no question that computing has dramatically changed society, and has furthered humanity in ways that were hard to foresee at its onset. Many factors have contributed to its unfettered success including the adoption of Boolean logic and algorithmic thinking, the invention of the instruction set machine, and the advent of the semiconductor technology. The latter offered us a semi-perfect switch device and effective ways of storing data. All these factors have led to an amazing run of almost 7 decades. Over time, the quest for ever higher performance in the presence of power and energy limitations have forced us to make major changes to how the processors were architected and operated—such as the introduction of concurrency, the adoption of co-processors and accelerators, or the adoption of ever more complex memory hierarchies. However, in essence the fundamentals remained unchanged

For a number of reasons, this model is at the verge of undergoing some major changes and challenges. On the one hand, the scaling model of semiconductors—commonly known as Moore's law—is running out of steam, hence depriving us from a convenient means in improving computational performance, density and efficiency. On the other hand, the nature of computation itself is changing with data rather than algorithm taking primacy. Both these trends force us to reflect on some of the foundational concepts that have driven computation for such a long period. In an abundance of data, statistical distributions become more relevant than deterministic answers. Many perceptual tasks related to human-world interaction fall under the same class. Learning-based programming approaches are gaining rapid interest and influence. Simultaneously the lack of "the perfect switch", as well as the high-variability of nanoscale devices operating under high energy-efficiency (that is, low-voltage) makes deterministic computing an extremely expensive if at all possible undertaking.

All of this has made researchers explore novel computational models that are "approximate" in nature. This means that errors and approximations are becoming acceptable as long as the outcomes have a well-defined statistical behavior. A number of approaches have been identified and are being actively pursued under different headers such as approximate computing, statistical computing and stochastic processing. In this book, the authors investigate how errors caused by variation (especially those caused by timing) can be exposed to the software layers, and how they can be mitigated using a range of techniques and methods to reduce their impact. The document provides a clear insight of what is possible through pure software intervention.

This book is at the forefront of what is to come at the frontiers in the new age of computation. As such, I heartily recommend it as a great intellectual effort and a superb read.

Jan M. Rabaey

# Preface

Variation in performance and power across manufactured parts and their operating conditions is an accepted reality in modern microelectronic manufacturing processes with geometries in nanometer scales. This book views such variations both as a challenge as well an opportunity to rearchitect the hardware/software interface that provides more resilient system architectures. We start with an examination of how variability manifests itself across various levels of microelectronic systems. We examine various mechanisms designers use, and can use, to combat negative effects of variability.

This book attempts a comprehensive look at the entire software/hardware stack and system architecture in order to devise effective strategies to address microelectronic variability. First, we review the key concepts on *timing errors* caused by various variability sources. We use a two-pronged strategy to mitigate such errors by jointly exposing hardware variations to the software and by exploiting flexibility made possible by parallel processing. We consider methods to predict and prevent, detect and correct, and finally conditions under which such errors can be accepted. For each of these methods, our work spans defining and measuring the notion of error tolerance at various levels, from instructions to procedures to parallel programs. These measures essentially capture the likelihood of errors and associated cost of error correction at different levels. The result is a design platform that enables us to combine these methods that enable detection and correction of erroneous results within a defined criterion for acceptable errors using a notion of memoization across the hardware/software interface. Pursuing this strategy, we develop a set of software techniques and microarchitecture optimizations for improving cost and scale of these methods in massively parallel computing units, such as general-purpose graphics processing units (GP-GPUs), clustered many-core architectures, and field-programmable gate array (FPGA) accelerators.

Our results show that parallel architectures and use of parallelism in general provides the best means to combat and exploit variability. Using such programmable parallel accelerator architectures, we show how system designers can coordinate propagation of error information and its effects along with new techniques for memoization and memristive associative memory. This book naturally leads to use of these techniques into emerging area of approximate computing, and how these can be used in building resilient and efficient computing systems.

Berkeley, USA — Abbas Rahimi
Zürich, Switzerland — Luca Benini
San Diego, USA — Rajesh K. Gupta
January 2017

# Contents

# Chapter 1
# Introduction

**Chapter Summary** Variation in performance and power consumption is a common phenomenon in semiconductor manufacturing. What makes it particularly challenging, however, is its effect on manufacturing of devices as these scale down to near atomic scale feature dimensions. Any variation in dimensions, doping, etc, has a large effect on the resulting device and circuit behavior. To address this variation, designers resort to design **guardbands**. These guardbands are increasing rapidly and eventually obliterating any gains due to device scaling. As a consequence, reduction of design guardbands in design has become an important research challenge with recent results that recover a part of these guardbands through circuit-level changes. We begin by examining sources of variability in integrated circuits and conclude with an outline of the entire book.

## 1.1 Sources of Variability

Broadly speaking, there are three physical types of variations: **(i) Spatial variability**: Process variations cause static variations in critical dimension, channel length (L), and threshold voltage ($V_{th}$) of devices due to dopant fluctuations and sub-wavelength lithography. These variations manifest themselves as die-to-die (D2D) and within-die (WID) variations [1]. D2D variations affect all devices on a die equally, whereas WID variations induce different characteristics for each device. **(ii) Temporal variability**: Aging and wearout mechanisms cause slow temporal degradation in devices reliability. Device aging mechanisms are induced by negative bias temperature instability (NBTI), positive bias temperature instability, electromigration, time-dependent dielectric breakdown, gate oxide integrity, thermal cycling, and hot carrier injection [2]. **(iii) Dynamic variability**: Environmental variations in ambient condition are caused by fluctuations in operating temperature and supply voltage droops. Voltage droops result from abrupt changes in the switching activity, inducing large current transients in the power delivery system (dI/dt voltage drops), and contain high-frequency and low-frequency components which occur locally as well as globally across the die [3]. On the other hand, temperature variations occur at a relatively slow time scale with local hot spots on the die, depending on environmental, and workload



A. Rahimi et al., *From Variability Tolerance to Approximate Computing in Parallel Integrated Architectures and Accelerators*,
DOI 10.1007/978-3-319-53768-9_1

conditions [4]. The origins of variability include time-independent DC component (process variations), slow-varying low-frequency components (aging and temperature), and fast-changing high-frequency components (voltage droops). The variations are expected to be worse with technology scaling [5].

Spatial parameter variations in the device geometries in conjunction with temporal degradation and undesirable fluctuations in the operating condition may prevent circuit from meeting the performance and power constraints. The most immediate manifestations of variability are in path delay (therefore, performance) and power variations. Sequential elements are connected at the end of the paths to hold the circuit state. Path delay variations cause violation of timing specification resulting in circuit-level *timing errors* that could lead to an invalid state being stored in the sequential element. This could result in a malfunction of the digital system. Synchronous circuit designers commonly handle the timing errors by adding safety timing margins to the voltage and/or the clock frequency as guardband. This practice leads to overly conservative designs. Currently, the guardbands tend to accumulate as design closure is performed using a multi-corner analysis, with an increasing number of corners [1, 6, 7]. As a result, the impact of guardbanding on the key design metrics (power, performance, and area) has been steadily increasing with technology scaling [5], leading to loss of operational efficiency and increased costs due to overdesign. Power variability is also challenging, for instance 13× variation in the sleep power across ten instances of ARM Cortex M3 core was observed over a temperature range of 22–60 °C [8]. This thesis focuses instead on the path delay variation and its manifestation as timing errors. We identify the timing error as the most threatening manifestation of variability and investigate various means to address it. We begin with a quantitative feel of the extent of variation currently seen in manufactured devices. Section 1.2 covers the delay variation in details.

## 1.2 Delay Variation

For an Intel 80-core processor in 65 nm, Fig. 1.1 shows the WID core-to-core maximum frequency (Fmax) variations for each of the 80 cores. The measurements have been done at a fixed operating temperature of 50 °C with three operating voltages: 1.2, 0.9, and 0.8 V. At the nominal voltage of 1.2 V, the fastest core displays the Fmax of 7.3 GHz while in the same die the slowest core can work with the Fmax of 5.7 GHz resulting in 28% WID clock frequency variation. Figure 1.2 illustrates the delay distribution of the 80 cores for the same operating conditions [9]. The single die with 80 cores exhibits an increasing value of $\sigma/\mu$ for lower voltages: 5.93, 6.37, and 8.64% for 1.2, 0.9, and 0.8 V, respectively. Lowering the voltage from the nominal 1.2 to 0.8 V, increases the critical paths variability ($\sigma/\mu$) by 45% [9].

Voltage overscaling (VOS) [10] and working at near-threshold (NT) voltage [11] have become popular approaches for building energy-efficient digital circuits. Operating at low voltages ($V_{DD} \leq 0.5$ V) unfortunately exacerbates the effects of delay variations [10, 12–15]. This indicates the importance of variability awareness at

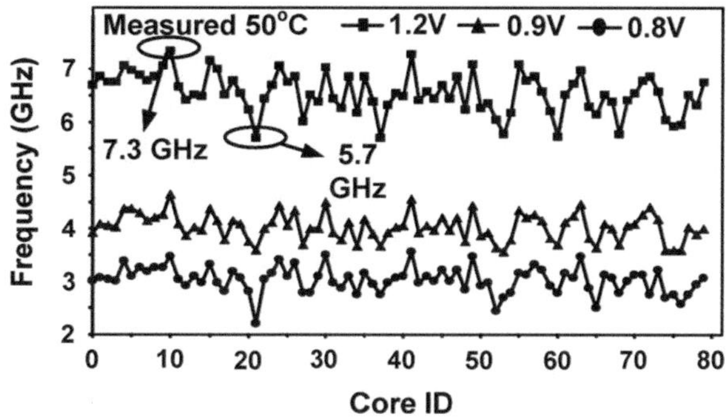


**Fig. 1.1** WID core-to-core maximum clock frequency variation for 80 cores on a single chip [9]

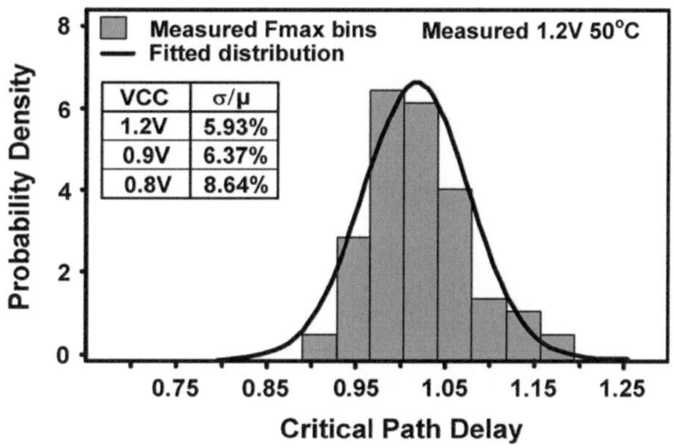


**Fig. 1.2** Critical path delay distribution and its coefficient of variation ($\sigma/\mu$) for 80 cores on a single chip [9]

lower operating voltages, where the delay uncertainty is further increased. The WID delay measurement for a 45 nm SIMD processor shows that reducing $V_{DD}$ from 1.0 to 0.53 V increases the delay variation by 6× [15]. Figure 1.3 shows the normalized gate delay variation due to process variations as a function of $V_{DD}$ [12]. Working at near-threshold voltage of 400 mV increases the performance variability by 5× compared to 1.3× at the nominal operating voltage. It is then clear that for logic working at near-threshold voltages, the statistical WID variation in the voltage threshold ($V_{th}$) plays an important role in determining the path delay. $V_{th}$ variations result mainly from random fluctuations in the number of dopant atoms in the transistor channels [13]. Considering dynamic sources of variations, including temperature fluctuations, and voltage droops results in a total performance variability of 20× [12].

Given such a growing increase in performance variability, design methods are needed to make a design resilient to timing errors, especially for circuits operating at low voltages where the effect of delay uncertainty is pronounced. The effects of the static process variations can sometimes be mitigated through binning or by post-silicon tuning during test time, while the dynamic variations manifest themselves on the field as a function of time and environment, and therefore cannot be compensated by *one-time* pre-silicon and post-silicon tuning techniques. Consequently, accurate design time analysis coupled with efficient runtime techniques are required to overcome the variability challenges.

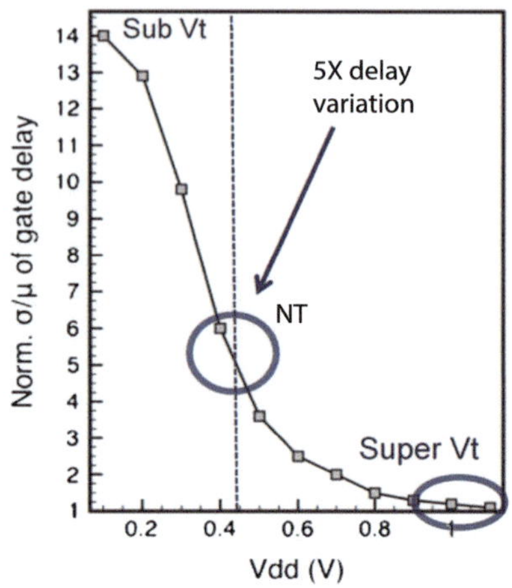


**Fig. 1.3** Impact of voltage scaling on gate delay variation due to process variation [12]

## 1.3 Book Organization

This book grew out of a doctoral dissertation at the University of California, San Diego [16]. This book focuses on timing errors caused by various sources of variations at different levels. We devise methods to mitigate such errors by jointly **exposing hardware variations to the software** and by **exploiting parallel processing**. We investigate methods to *predict and prevent*, *detect and correct*, and finally conditions under which errors can be *accepted*. We classify our proposed methods into a conceptual Y-chart shown in Fig. 1.4.

The Y-chart in Fig. 1.4 groups these methods to address variability into three parts based on *when* and *how* the timing errors should be manipulated. These three parts of the Y-chart are on radial axes. The first axis describes mainly *design time* approaches for predicting and preventing timing errors. The second axis focuses on *runtime* approaches for detecting and correcting timing errors, while the third axis accepts timing errors if possible. Further, we combine these two axes to devise a new joint method of detecting and correcting with accepting errors. Each part is divided into levels of abstraction, using concentric rings. Every abstraction level determines at *which* level of the computing stack the approaches can be applied: *circuit*, *architecture*, and *software*. At the top-level outer ring, we consider approaches applicable to software level; at the lower levels inner rings, we refine approaches into finer architecture, and circuit implementations.

Thus, Fig. 1.4 puts our work in perspective, with the three main axes defining the three separate methodological approaches. For each of these approaches, our work spanned defining and measuring the notion of error tolerance, from instruction set architecture (ISA) to procedures to parallel programs. These measures essentially capture the likelihood of errors and associated cost of error correction at different levels.

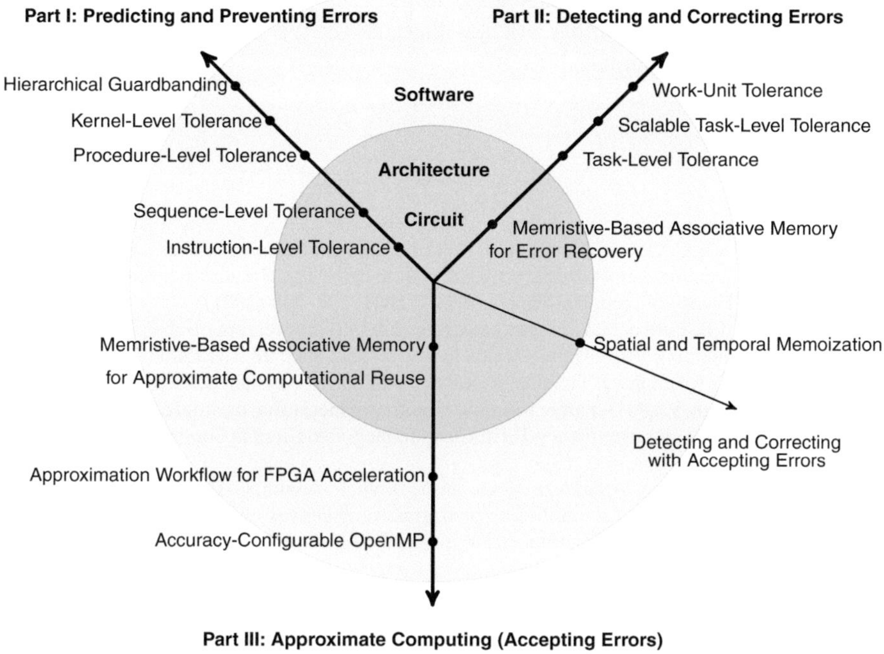


**Fig. 1.4** Taxonomy of timing error tolerance in this book: abstractions versus approaches

Next natural step is to see the possibility and consequences of relaxing the notion of accuracy and precision in computation. We focus on parallel programming, runtime environment, parallel integrated architecture, and accelerators to support controlled "approximate computing." That is, ensuring safety of error mitigation methods through a set of rules verified by a combination of design-time and runtime constraints. The goal is to deliver functionality within specified quality guarantees. The result is a new joint method of detecting and correcting with accepting errors across the hardware/software interface using memoization techniques spatially or across time (i.e., spatial or temporal reuse of computation). We accordingly devise an arsenal of software techniques and microarchitecture optimizations for improving cost and scale of these methods in massively parallel computing units, such as general-purpose graphics processing units (GP-GPUs), clustered many-core architectures, and field-programmable gate array (FPGA) accelerators. The main focus of our attention for the data-level parallelism is on single instruction, multiple data (SIMD) and GP-GPU architectures, and for the task-level parallelism is on the shared-memory processor clusters. We find that parallel architectures and parallelism in general provide the best means to combat and exploit variability to design resilient and efficient systems. Using such programmable parallel accelerator architectures, we show how system designers can coordinate propagation of error information and its effects along with new techniques for memoization and memristive associative memory. This discussion naturally leads to use of these techniques into emerging area of approximate computing, and how these can be used in building resilient and efficient computing systems.

## References


1. K.A. Bowman, S.G. Duvall, J.D. Meindl, Impact of die-to-die and within-die parameter fluctuations on the maximum clock frequency distribution, in IEEE International Solid-State Circuits Conference, Digest of Technical Papers. ISSCC **2001**, 278–279 (2001)
2. X. Li, J. Qin, J.B. Bernstein, Compact modeling of MOSFET wearout mechanisms for circuit-reliability simulation. IEEE Trans. Device Mater. Reliab. **8**(1), 98–121 (2008)
3. K. Bowman, C. Tokunaga, J. Tschanz, A. Raychowdhury, M. Khellah, B. Geuskens, S.-L. Lu, P. Aseron, T. Karnik, V. De, Dynamic variation monitor for measuring the impact of voltage droops on microprocessor clock frequency. IEEE Custom Integrated Circuits Conference (CICC) **2010**, 1–4 (2010)
4. S. Murali, A. Mutapcic, D. Atienza, R. Gupta, S. Boyd, L. Benini, G. De Micheli, Temperature control of high-performance multi-core platforms using convex optimization, in *Proceedings of the Conference on Design, Automation and Test in Europe*, DATE '08, ACM, New York, NY, USA (2008), pp. 110–115
5. The ITRS website. http://www.itrs.net/Links/2011ITRS/Home2011.htm
6. S. Borkar, T. Karnik, S. Narendra, J. Tschanz, A. Keshavarzi, V. De, Parameter variations and impact on circuits and microarchitecture. Proceedings of Design Automation Conference **2003**, 338–342 (2003)
7. T. Austin, V. Bertacco, D. Blaauw, T. Mudge, Opportunities and challenges for better than worst-case design, in *Proceedings of the 2005 Asia and South Pacific Design Automation Conference*, ASP-DAC '05, ACM, New York, NY, USA (2005), pp. 2–7
8. L. Wanner, R. Balani, S. Zahedi, C. Apte, P. Gupta, M. Srivastava, Variability-aware duty cycle scheduling in long running embedded sensing systems, in Design. Automation Test in Europe Conference Exhibition (DATE) **2011**, 1–6 (2011)
9. S. Dighe, S.R. Vangal, P. Aseron, S. Kumar, T. Jacob, K.A. Bowman, J. Howard, J. Tschanz, V. Erraguntla, N. Borkar, V.K. De, S. Borkar, Within-die variation-aware dynamic-voltage-frequency-scaling with optimal core allocation and thread hopping for the 80-core teraflops processor. IEEE J. Solid-State Circuits **46**(1), 184–193 (2011)
10. D. Jeon, M. Seok, Z. Zhang, D. Blaauw, D. Sylvester, Design methodology for voltage-overscaled ultra-low-power systems. IEEE Trans. Circuits Syst. II Express. Briefs **59**(12), 952–956 (2012)
11. B. Zhai, R.G. Dreslinski, D. Blaauw, T. Mudge, D. Sylvester, Energy efficient near-threshold chip multi-processing. ACM/IEEE International Symposium on Low Power Electronics and Design (ISLPED) **2007**, 32–37 (2007)
12. R.G. Dreslinski, M. Wieckowski, D. Blaauw, D. Sylvester, T.N. Mudge, Near-threshold computing: reclaiming moore's law through energy efficient integrated circuits. Proc. IEEE **98**(2), 253–266 (2010)
13. R. Rithe, S. Chou, J. Gu, A. Wang, S. Datla, G. Gammie, D. Buss, A. Chandrakasan, The effect of random dopant fluctuations on logic timing at low voltage. IEEE Trans. Very Large Scale Integr. (VLSI) Syst., **20**(5), 911–924
14. M.R. Kakoee, I. Loi, L. Benini, Variation-tolerant architecture for ultra low power shared-l1 processor clusters. IEEE Trans. Circuits Syst. II Express. Briefs **59**(12), 927–931 (2012)

15. R. Pawlowski, E. Krimer, J. Crop, J. Postman, N. Moezzi-Madani, M. Erez, P. Chiang, A 530mV 10-lane SIMD processor with variation resiliency in 45nm SOI. IEEE International Solid-State Circuits Conference Digest of Technical Papers (ISSCC) **2012**, 492–494 (2012)
16. Abbas Rahimi, "From Variability-Tolerance to Approximate Computing in Parallel Computing Architectures," Ph.D. Dissertation, Department of Computer Science and Engineering, University of California, San Diego, CA, (2015)

# Part I
# Predicting and Preventing Errors

In this part, we explore approaches to reduce the excessive guardband and enable better than worst-case design while avoiding the timing errors [124–128]. These methods typically use characterization metrics and rely upon modeling that derives rules for simultaneous guardband reduction and error prevention. We show how these methods can be implemented at different abstraction levels, from instructions to kernels.We first characterize instructions for the effect of circuit timing errors on tolerance of individual (Chap. 2), or streams (Chap. 3) of instructions when executing on a single core architecture. Raising further the level of abstraction, procedure-level tolerance (Chap. 4) exposes the effect of dynamic variations to procedure calls for use in software preventive actions. Such exposure leads to a low overhead procedure hopping technique for a tightly coupled processor clusters with 16 cores. This is even more challenging in GP-GPUs and other many-core accelerators where the effect of these variations is not uniformly spread across over thousands processing elements: some are affected more (hence less reliable) than others. In this regard, we propose two methods suitable for GP-GPUs that adaptively predict the delay variations and react accordingly to prevent the timing errors. This first method is an adaptive very long instruction word (VLIW) assignment which is described in Chap. 5. We devise an adaptive compiler method that equalizes the expected lifetime of each processing element by regenerating aging-aware healthy kernels that respond to the specific health state of GP-GPUs. This aging-aware compiler periodically exposes the inherent idleness in VLIW slots and guides its distribution that does matter for the aging. This reallocation mitigates the impacts on lifetime uncertainty and unbalancing among the processing elements. The second method adaptively avoids PVT and aging (PVTA) induced timing errors. Using a model based on supervised learning and PVTA monitoring circuits, we propose hierarchically focused guardbanding (HFG) and demonstrate its effectiveness on GP-GPU architecture at two granularities of observation and adaptation: (i) fine-grained instruction-level; and (ii) coarse-grained kernel-level. Chapter 6 describes HFG in details.

# Chapter 2
# Instruction-Level Tolerance

**Abstract** Microprocessors manufactured in nanometer processes are beginning to see variation in timing performance of individual instructions. This chapter considers challenges and opportunities in identifying this variation and methods to combat it for predicting and preventing the timing errors in single-core architectures. We start from instruction-level which is the finest granularity to present the processor functionality. We introduce the notion of instruction-level vulnerability (ILV) to parameterize this variation and use it for architectural and compiler optimizations. To compute ILV, we quantify the effect of voltage and temperature variations on the performance and power of a 32-bit RISC in-order processor in 65 nm TSMC technology at the level of individual instructions. Results show 3.4 ns (68 fanout of 4 or 68FO4) delay variation and 26.7× power variation among instructions, and across extreme corners. Our analysis shows that ILV is not uniform across the instruction set. In fact, ILV data partitions instructions into three equivalence classes. Based on this classification, we show how low-overhead monitors and adaptive clocking techniques can be used to enhance performance by a factor of 1.1×–5.5×.

## 2.1 Introduction

Designers commonly use conservative guardbands into the operating frequency and voltage to handle these variations to ensure error-free operation within the presence of worse case dynamic variations over circuit lifetime that leads to loss of operational efficiency. An alternative is to use sensor circuits to detect dynamic variations coupled with an adaptive recovery methods for quick on-line error detection and compensation.

Further progress in this area requires a careful analysis of the effect of variations on individual instructions. Here we advance the state of the art through following three means:

1. We analyze the effect of a full range of voltage and temperature variations on the performance and power of the 32-bit in-order RISC LEON-3 [1] processor (Sect. 2.2).

 
A. Rahimi et al., *From Variability Tolerance to Approximate Computing in Parallel Integrated Architectures and Accelerators*,
DOI 10.1007/978-3-319-53768-9_2

2. We introduce the notion of instruction-level vulnerability (ILV) to characterize tolerance of individual instruction to dynamic variations. ILV exposes variation and its effects to the software stack for use in architectural/compiler optimizations. Our results show that ILV is not uniform across the instruction set (Sects. 2.3 and 2.4).
3. Using ILV data, we show the effectiveness of a minimally intrusive and cost-effective fine-grained technique to mitigate the dynamic variations that achieves up to 5.5$\times$ performance improvement in comparison to the traditional worst-case design.

## 2.2 Effect of Operating Conditions

We analyze the effect of operating conditions on the performance and power of the LEON-3 [1] processor compliant with the SPARC V8 architecture. Specifically, we used a temperature range of –40–125 °C, and a voltage range of 0.72–1.1 V. Figure 2.1 shows how the critical path of the processor varies across corners. The higher voltage results in the shorter critical path, while the lower temperature leads to a higher delay in the low-voltage region (voltage $\leq$0.9 V), since MOSFET drain current decreases when the temperature is decreased in the deep submicron technologies [2]. These operating condition (hence dynamic) variations cause the critical path delay to increase by a factor of 6.1$\times$ when the operating condition is varied from the one corner to the other. Consequently, a large conservative guardband into the operating frequency is needed to ensure the error-free operation in presence of the dynamic variations.

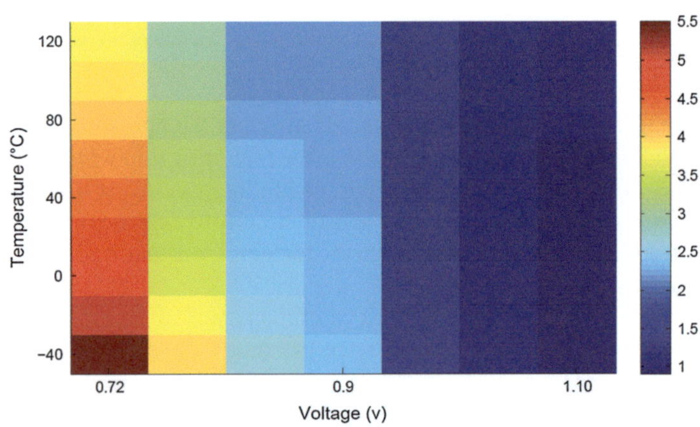


**Fig. 2.1** Effect of voltage and temperature variations on the critical path (ns)

## 2.3 Delay Variation Among Pipeline Stages

We now evaluate the critical paths of each pipeline stage for a given cycle time, while changing the operating conditions. Figure 2.2 shows the number of failed paths with a negative slack for each parallel pipeline stages across three corners. The cycle time is set at 0.85 ns, and voltage varies from 0.72 to 0.88 V, and then to 1.10 V at a constant temperature of 125 °C. As shown in Fig. 2.2, most of the failed paths lie in the execute and memory stages in all three operating voltages. On the other hand, each of the fetch, decode, and register access stages contains less than 40 K failed paths. Furthermore, there is a relatively small fluctuation in their number of critical paths across voltage variations for these stages. Quantitatively, the memory stage at operating voltage of 0.72 V has 1.3×, 1.8×, 3.8× more critical paths in comparison to the execute, write back, and decode stages, respectively. Memory stage at operating voltage of 1.10 V also faces 1.4×, 1.9× more critical paths when the voltage drops to 0.88, 0.72 V, respectively.

Similarly, in Fig. 2.3 the temperature of processor is varied from –40 to 125 °C at a constant voltage of 1.1 V. As a result, there are no failed paths in the fetch stage when the temperature is varied, and only a small number of failed paths are found in the write back stage at the highest temperature. On the other hand, similar to Fig. 2.2, many paths fail within the execute and memory stages. The execute and memory parts of the processor are not only very sensitive to voltage and temperature variations, but also exhibit a large number of critical paths in comparison to the rest of processor. *Therefore, we would anticipate that the instructions that significantly exercise the execute and memory stages are likely to be more vulnerable to voltage and temperature variations.*

Let us now examine the situation of all paths through the processor under different operating condition and frequency. The Y-axis of Fig. 2.4 shows the proportion of failed paths to nonfailed paths for three corners. We observe that this proportion of failed paths suddenly drops below a certain threshold while the clock is finely scaled with a resolution of 0.01 ns. For instance, the proportion falls below 0.5 with only 0.06 ns clock scaling at (1.10 V, 0 °C); in the other words, the number of nonfailed paths is twice as many as those which fail. Alternatively, the number of nonfailed

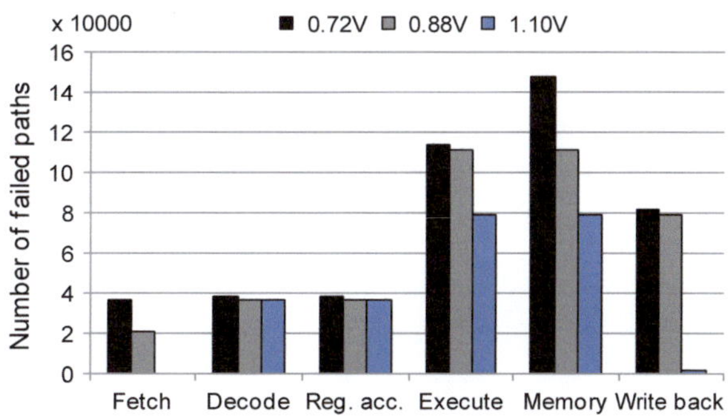


**Fig. 2.2** Effect of voltage variation on the pipeline stages at 125 °C

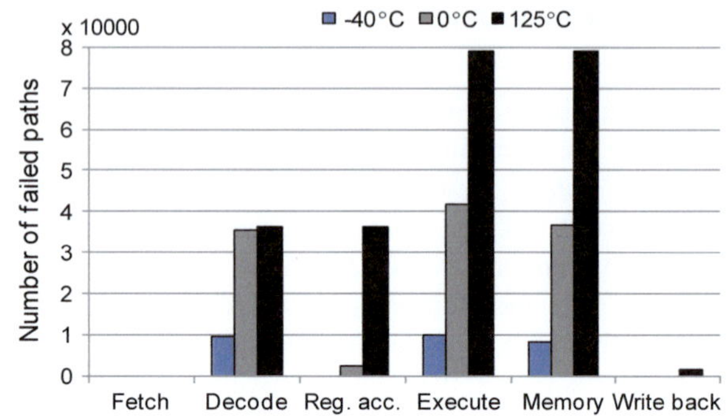


**Fig. 2.3** Effect of temperature variation on the number of failed paths among the pipeline stages at 1.10 V

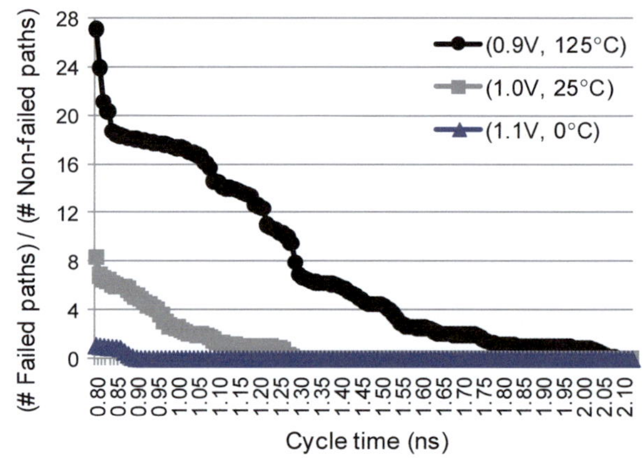


**Fig. 2.4** The proportion of failed paths to nonfailed paths versus clock

paths is doubled when the cycle time is increased for 0.3 ns at (0.9 V, 125 °C). These provide an opportunity for an error-free running of some instructions that will not activate those failed paths.

From the previous analysis, we see that instructions will have different levels of vulnerability to variations depending on the way they exercise the nonuniform critical paths across the various pipeline stages. To capture this phenomenon, we define the concept of instruction-level vulnerability to dynamic variations. The classification of instructions is a valuable mechanism to alleviate the guardbanding and improving performance: (i) within a fixed corner, by acquiring the knowledge about which class of instructions is running, the processor can adapt the guardbanding accordingly, without any need for the intrusive variability sensor/observer; (ii) across every corner, processor can adjust its guardbanding for all class of instructions by using a low-overhead variability observer, e.g., phase locked loop (PLL) [3], and ring oscillators (RO) [4].

## 2.4 Instruction Characterization Methodology and Experimental Results

We use integer pipeline of LEON-3 processor with hardware multiplier/divider units as well as the instruction/data caches to characterize instructions. First, we synthesized the open-source VHDL code of LEON-3 with the TSMC 65 nm technology library (general purpose process) to generate gate-level netlist. The signoff stage for accurate analysis of the operating conditions has been made with Synopsys PrimeTime, using its voltage-temperature scaling features for the composite current source approach of modeling cell behavior. Mentor Graphics' ModelSim is also used for detail gate-level simulations.

### *2.4.1 Gate-Level Simulation*

In the gate-level simulation, for each individual instruction, we apply the Monte Carlo method to observe instruction behavior. To accurately exercise each instruction, we use a normal distribution for the sources, destination, and immediate operands. A large sample of the SPARC ISA is evaluated, including the logical/arithmetic instructions, memory access instructions (load/store), multiply/divide instructions. To quantify the ILV to voltage and temperature variations, we define the **probability of failure (PoF)** for each instruction$_i$ in Eq. 2.1, where $N_i$ is the total number of clock cycles in Monte Carlo simulation which takes to execute instruction$_i$ with random operands; and Violation$_j$ indicates whether there is a violated stage at clock cycle$_j$ or not.

$$\text{PoF}i = \frac{1}{Ni}\sum_{j=1}^{\text{N}i} \text{Violation}j$$

$$\text{Violation}j = \begin{cases} 1 & \text{If any stage violates at cycle} j \\ 0 & \text{otherwise} \end{cases} \tag{2.1}$$

In other words, PoF$_i$ defines as the total number of violated cycles over the total simulated cycles for the instruction$_i$. If any of the analyzed stages have one or more violated flip-flop at clock cycle$_j$, we consider that stage as a violated stage at cycle$_j$. Intuitively, if instruction$_i$ runs without any violated path, PoF$_i$ is 0; on the other hand, PoF$_i$ is 1 if instruction$_i$ faces at least one violated path in any stage, in every cycle.

**Table 2.1** Probability of failure of ISA at 1.1 and 1.0 V, while varying temperature and frequency

| Corners | | (1.1V, -40°C) | | | (1.1V, 0°C) | | | (1.1V, 125°C) | | | | | | (1.0V, 25°C) | | | | |
|---|---|---|---|---|---|---|---|---|---|---|---|---|---|---|---|---|---|---|
| Cycle time (ns) | | 0.74 | 0.76 | 0.78 | 0.74 | 0.76 | 0.78 | 0.80 | 0.82 | 0.84 | 0.86 | 0.88 | 0.90 | 1.08 | 1.10 | 1.12 | 1.14 | 1.22 |
| Logical & Arithmetic | add | 1 | 0 | 0 | 1 | 0 | 0 | 1 | 0 | 0 | 0 | 0 | 0 | 1 | 0 | 0 | 0 | 0 |
| | and | 1 | 0 | 0 | 1 | 0 | 0 | 1 | 0 | 0 | 0 | 0 | 0 | 1 | 0 | 0 | 0 | 0 |
| | or | 1 | 0 | 0 | 1 | 0 | 0 | 1 | 0 | 0 | 0 | 0 | 0 | 1 | 0 | 0 | 0 | 0 |
| | sll | 1 | 0 | 0 | 1 | 0 | 0 | 1 | 0 | 0 | 0 | 0 | 0 | 1 | 0 | 0 | 0 | 0 |
| | sra | 1 | 0 | 0 | 1 | 0 | 0 | 1 | 0 | 0 | 0 | 0 | 0 | 1 | 0 | 0 | 0 | 0 |
| | srl | 1 | 0 | 0 | 1 | 0 | 0 | 1 | 0 | 0 | 0 | 0 | 0 | 1 | 0 | 0 | 0 | 0 |
| | sub | 1 | 0 | 0 | 1 | 0 | 0 | 1 | 0 | 0 | 0 | 0 | 0 | 1 | 0 | 0 | 0 | 0 |
| | xnor | 1 | 0 | 0 | 1 | 0 | 0 | 1 | 0 | 0 | 0 | 0 | 0 | 1 | 0 | 0 | 0 | 0 |
| | xor | 1 | 0 | 0 | 1 | 0 | 0 | 1 | 0 | 0 | 0 | 0 | 0 | 1 | 0 | 0 | 0 | 0 |
| Mem | load | 1 | 0 | 0 | 1 | 0 | 0 | 1 | 0 | 0 | 0 | 0 | 0 | 1 | 0.786 | 0 | 0 | 0 |
| | store | 1 | 0 | 0 | 1 | 0 | 0 | 1 | 0 | 0 | 0 | 0 | 0 | 1 | 0.814 | 0 | 0 | 0 |
| Mul. &Div | mul | 1 | 0 | 0 | 1 | 0.967 | 0 | 1 | 0.042 | 0.015 | 0.012 | 0.002 | 0 | 1 | 0.998 | 0.976 | 0.074 | 0 |
| | div | 1 | 0.837 | 0 | 1 | 0.948 | 0 | 1 | 0.991 | 0.991 | 0.984 | 0.984 | 0 | 1 | 0.964 | 0.993 | 0.990 | 0 |

### 2.4.2 *Instruction-Level Delay Variability*

Tables 2.1 and 2.2 summarize the PoF of each evaluated instruction across various corners. We finely change the clock cycle to observe the paths failure for every exercised instruction, and then consequently evaluate its PoF. As shown, instructions exhibit a very wide range of delay under different operating conditions ranges from 0.76 to 4.16 ns. More precisely, the PoF values shown in tables evidence two important facts. First, for their vulnerability to variations, instructions are partitioned into three main classes: (i) the logical/arithmetic instructions, (ii) the memory instructions, and (iii) the multiply/divide instructions. The 1st class shows an abrupt behavior when the clock cycle is slightly varied. Its PoF switches from 1 to 0 with a slight increase in the cycle time (0.02 ns) for every corner, mainly because the path distribution of the exercised part by this class is such that most of the paths have the same length, then we have a all-or-nothing effect, which implies that either all instructions within this class fail or all make it. The 2nd class, the memory instructions, needs much more relaxed cycle time to be able to survive across conditions. For instance, as shown in Table 2.2, only 0.04 ns more guardbanding on the cycle time of the 1st class instruction can guarantee the error-free execution of the memory instructions while they are experiencing 40 °C temperature fluctuation. The 3rd class is the multiply/divide instructions which need higher guardbanding in comparison to the 1st class instruction, ranges from 0.02 ns at (1.1 V, –40 °C) to 0.30 ns at (0.72 V, 125 °C). Since this class highly exercises the execution unit,[1] it has a higher PoF in comparison with the rest of classes in the same clock cycle, for every corner.

Further, based on these results, we can define an adaptive clock cycle for each class of instructions to mitigate the conservative guardbanding, not only within a fixed process corner, but also across corners. All instruction classes act similarly across the wide range of operating conditions: as the cycle time increases gradually, the PoF becomes 0, firstly for the 1st class, then for the 2nd class, and finally for the 3rd

[1]Moreover, 64–82% (depends on the corner) of the failed paths in the execution stage lie in the hardware multiplier and divider units.

**Table 2.2** Probability of failure of ISA at constant voltage 0.72V, while varying temperature and frequency

| Corners | | (0.72V, -40°C) | | | | (0.72V, 0°C) | | | | | (0.72V, 125°C) | | | | | | |
|---|---|---|---|---|---|---|---|---|---|---|---|---|---|---|---|---|---|
| Cycle time (ns) | | 4.10 | 4.12 | 4.14 | 4.16 | 3.58 | 3.60 | 3.62 | 3.64 | 3.66 | 2.88 | 2.90 | 2.92 | 2.94 | 2.98 | 3.00 | 3.20 |
| Logical & Arithmetic | add | 1 | 0 | 0 | 0 | 1 | 0 | 0 | 0 | 0 | 1 | 0 | 0 | 0 | 0 | 0 | 0 |
| | and | 1 | 0 | 0 | 0 | 1 | 0 | 0 | 0 | 0 | 1 | 0 | 0 | 0 | 0 | 0 | 0 |
| | or | 1 | 0 | 0 | 0 | 1 | 0 | 0 | 0 | 0 | 1 | 0 | 0 | 0 | 0 | 0 | 0 |
| | sll | 1 | 0 | 0 | 0 | 1 | 0 | 0 | 0 | 0 | 1 | 0 | 0 | 0 | 0 | 0 | 0 |
| | sra | 1 | 0 | 0 | 0 | 1 | 0 | 0 | 0 | 0 | 1 | 0 | 0 | 0 | 0 | 0 | 0 |
| | srl | 1 | 0 | 0 | 0 | 1 | 0 | 0 | 0 | 0 | 1 | 0 | 0 | 0 | 0 | 0 | 0 |
| | sub | 1 | 0 | 0 | 0 | 1 | 0 | 0 | 0 | 0 | 1 | 0 | 0 | 0 | 0 | 0 | 0 |
| | xnor | 1 | 0 | 0 | 0 | 1 | 0 | 0 | 0 | 0 | 1 | 0 | 0 | 0 | 0 | 0 | 0 |
| | xor | 1 | 0 | 0 | 0 | 1 | 0 | 0 | 0 | 0 | 1 | 0 | 0 | 0 | 0 | 0 | 0 |
| Mem | load | 1 | 0.823 | 0.823 | 0 | 1 | 0.823 | 0.823 | 0 | 0 | 1 | 0.823 | 0.823 | 0.823 | 0.796 | 0.796 | 0 |
| | store | 1 | 0.847 | 0.847 | 0 | 1 | 0.847 | 0.847 | 0 | 0 | 1 | 0.847 | 0.847 | 0.847 | 0.823 | 0.823 | 0 |
| Mul. &Div | mul | 1 | 0.995 | 0.995 | 0 | 1 | 0.996 | 0.994 | 0 | 0 | 1 | 0.998 | 0.997 | 0.996 | 0.996 | 0.996 | 0 |
| | div | 1 | 0.995 | 0.995 | 0 | 1 | 0.995 | 0.995 | 0.812 | 0 | 1 | 0.994 | 0.994 | 0.993 | 0.991 | 0.991 | 0 |

class. A processor can benefit from this classification by adapting its guardbanding for each class of instruction by acquiring the knowledge about which class of instructions is/will be running.

### *2.4.3 Less Intrusive Variation-Tolerant Technique*

All intrusive techniques [5–7] try to avoid timing failure for instructions that activate the critical paths by dynamically switching to two-cycle operation. These expensive, instruction by instruction timing adjustment techniques do not expose opportunity for further software-level optimizations especially for sequences and classes of instructions. Therefore, we could have an advanced dynamic clock speed adaptation technique, possibly compiler driven, which can quickly decide on the clock speed of the processor at a very fine-grained [8], just looking at the fetched instructions and keeping track of their entry into the stages, and at the same time monitoring the current corner with a low-overhead monitoring hardware [3, 4]. This technique not only provides great performance enhancement for processor, but also is a step forward toward a less intrusive circuit monitoring and cost-effective robust design.

Table 2.3 shows how a program consisting of various classes of instructions can benefit by this technique under different operating conditions: the performance improvement when processor runs a program only consists of specific classes, in comparison to the traditional worst-case design. For instance, at the typical operating condition (1.0 V, 25 °C) processor can decrease the cycle time form 4.16 ns (Table 2.2) to 1.22 ns (Table 2.1), and consequently achieves 3.4× speed improvement, when its running program consists of all three classes. It can further reduce the cycle time to 1.12 ns (3.7× speedup) when only the 1st, and 2nd classes of instructions are used in its program. As shown, the proposed solution can greatly achieve 1.1 × –5.5 × performance improvement depends on the type of instruction and the operating condition.

**Table 2.3** Performance improvement for different classes of instructions

| Vol. (V) | Temp. (°C) | 1st and 2nd class | 1st, 2nd, 3rd class |
|---|---|---|---|
| 1.10 | −40 | 5.5× | 5.3× |
| 1.10 | 0 | 5.5× | 5.3× |
| 1.10 | 125 | 5.1× | 4.6× |
| 1.00 | 25 | 3.7× | 3.4× |
| 0.88 | −40 | 3.9× | 3.7× |
| 0.88 | 0 | 3.9× | 3.7× |
| 0.88 | 125 | 3.9× | 3.5× |
| 0.72 | 0 | 1.1× | 1.1× |
| 0.72 | 125 | 1.3× | 1.3× |

### *2.4.4 Power Variability*

From delay variability of instructions, we examine now variation of power consumption across and within process corners. The total power consumption of the instruction classes under four operating conditions is shown in Fig. 2.5, when the cycle time is adjusted for each class accordingly, i.e., the best frequency for each class is applied. As a result, all three classes of instructions experience a wide range of total power variability (0.1–2.6 mW), 1.15× intra-corner power variation (across the three classes) due to exercising various parts of processor, and 26.7× inter-corner power variation, at maximum. This implies that ILV could potentially expose opportunity for further software-level optimizations for both performance and power.

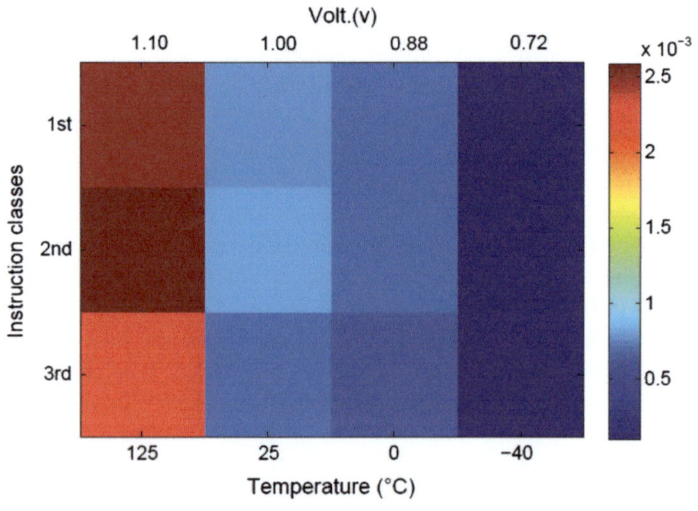


**Fig. 2.5** Intera- and inter-corner total power (W) variability of the instruction classes

## 2.5 Chapter Summary

The concept of instruction-level vulnerability to dynamic voltage and temperature variations is defined. Based on that, all exercised instructions in the integer pipeline of LEON-3 are partitioned into three classes for the full range of operating condition: (i) the logical and arithmetic instructions, (ii) the memory instructions, and (iii) the multiply and divide instructions. Using this classification in conjunction with less intrusive variability observers provides architectural/compiler optimizations a great opportunity to enhance processor performance by $1.1\times$–$5.5\times$, in TSMC 65 nm technology. It is also a step forward toward a low-overhead, efficient, and cost-effective robust design.

## References

1. Leon3. http://www.gaisler.com/cms/
2. R. Kumar, V. Kursun, Reversed temperature-dependent propagation delay characteristics in nanometer cmos circuits. IEEE Trans. Circuits Syst. II Express Briefs **53**(10), 1078–1082 (2006)
3. K. Kang, S.P. Park, K. Kim, K. Roy, On-chip variability sensor using phase-locked loop for detecting and correcting parametric timing failures. IEEE Trans. Very Large Scale Integr. (VLSI) Syst. **18**(2), 270–280 (2010)
4. M. Bhushan, A. Gattiker, M.B. Ketchen, K.K. Das, Ring oscillators for CMOS process tuning and variability control. IEEE Trans. Semicond. Manufact. **19**(1), 10–18 (2006)
5. D. Ernst, S. Das, S. Lee, D. Blaauw, T. Austin, T. Mudge, N.S. Kim, K. Flautner, Razor: circuit-level correction of timing errors for low-power operation. Micro, IEEE, **24**(6), 10–20 (2004)
6. K.A. Bowman, J.W. Tschanz, S.L. Lu, P.A. Aseron, M.M. Khellah, A. Raychowdhury, B.M. Geuskens, C. Tokunaga, C.B. Wilkerson, T. Karnik, V.K. De, A 45 nm resilient microprocessor core for dynamic variation tolerance. IEEE J. Solid-State Circuits **46**(1), 194–208 (2011)
7. D. Bull, S. Das, K. Shivshankar, G. Dasika, K. Flautner, D. Blaauw, A power-efficient 32b ARM ISA processor using timing-error detection and correction for transient-error tolerance and adaptation to PVT variation, in *2010 IEEE International Solid-State Circuits Conference Digest of Technical Papers (ISSCC)* (2010), pp. 284–285
8. J. Tschanz, N.S. Kim, S. Dighe, J. Howard, G. Ruhl, S. Vangal, S. Narendra, Y. Hoskote, H. Wilson, C. Lam, M. Shuman, C. Tokunaga, D. Somasekhar, S. Tang, D. Finan, T. Karnik, N. Borkar, N. Kurd, V. De, Adaptive frequency and biasing techniques for tolerance to dynamic temperature-voltage variations and aging. In *IEEE International Solid-State Circuits Conference, 2007. ISSCC 2007. Digest of Technical Papers* (2007), pp. 292–604

# Chapter 3
# Sequence-Level Tolerance

**Abstract** This chapter presents a method for predicting and preventing the timing errors for a sequence of instruction in single-core architectures. We introduce the notion of sequence-level vulnerability (SLV) that utilizes circuit-level vulnerability for constructing high-level software knowledge as metadata. In effect, the SLV metadata partitions sequences of integer SPARC instructions into two equivalence classes to enable an adaptive guardbanding technique to adapt the clock frequency simultaneously for dynamic voltage and temperature variations, as well as adapting to the different classes of the instruction sequences. The proposed technique achieves on an average 1.6× speedup for error-intolerant applications compared to recent work (Hoang, Exploring circuit timing-aware language and compilation, 2011, [1]). However, in reality, applications exhibit varying degrees of tolerance to error in computations. This chapter also proposes an adaptive guardbanding technique for error-tolerant (probabilistic) applications where application execution does not assume an error-free execution hardware. The proposed technique leverages a combination of accurate design time analysis and a minimally intrusive runtime technique to mitigate process, voltage, and temperature (PVT) variations for a near-zero area overhead. We demonstrate our approach on a 32-bit in-order RISC processor with full post placement and routing (P&R) layout results in TSMC 45 nm technology. The adaptive guardbanding technique eliminates traditional guardbands on clock frequency using information from PVT variations and application-specific requirements on computational accuracy. For probabilistic applications, the adaptive technique guarantees the error-free operation of a set of paths of the processor that always require correct timing (vulnerable paths) while reducing the cost of guardbanding for the rest of the paths (invulnerable paths). This increases the throughput of probabilistic applications upto 1.9× over the traditional worst-case design. The proposed technique has 0.022% area overhead, and imposes only 0.034 and 0.031% total power overhead for intolerant and probabilistic applications respectively.

## 3.1 Introduction

Several recent efforts have focused on measures to mitigate variability through innovations in circuit-level designs. These methods strive to achieve instruction



A. Rahimi et al., *From Variability Tolerance to Approximate Computing in Parallel Integrated Architectures and Accelerators*,
DOI 10.1007/978-3-319-53768-9_3

executions *exactly* as specified by the application programs. In contrast, **probabilistic** programs can exhibit enhanced error resilience at the application-level when multiple valid output values are permitted. Accurate design time analysis coupled with efficient runtime techniques are required to overcome the variability challenges. We propose a near-zero area overhead adaptive guardbanding technique to meet application-specific requirements on computational accuracy. This chapter makes the following contributions:

1. We present a method to relate low-level hardware vulnerability information obtained using accurate and practical variation-aware analysis to high-level knowledge in software. Our analysis flow considers the dynamic voltage and temperature as well as static process variations, and validates results on a full post P&R layout of a 32-bit in-order RISC processor.
2. We propose an adaptive guardbanding technique to dynamically adjust the cycle time to PVT variations and application-level computation accuracy. For probabilistic applications represented by multimedia benchmarks from MiBench [2] and MediaBench [3], the technique achieves up to 1.9× throughput improvement in comparison to the traditional worst-case design.
3. For error-intolerant applications, we introduce the notion of sequence-level vulnerability (SLV) to dynamic voltage and temperature variations. Our experimental results and analysis show that SLV is not uniform across sequences obtained from a large set of general purpose benchmarks [2–6]. Effectively, the SLV partitions sequences of integer SPARC instructions into two classes: ClassI, which only consists of the arithmetic/logical instructions; and ClassII, a mixture of all types of instructions. We also show the effectiveness of compiler technique to achieve a favorable mix of sequences. Using SLV enables the processor to achieve 1.6× average speedup for intolerant applications, compared to [1], by adapting the cycle time for dynamic variations and different instruction sequences. The minimally intrusive and cost-effective guardbanding in software greatly reduces the hardware cost with respect to the above-mentioned circuit techniques. Full layout results on TSMC 45 nm technology show that the proposed guardbanding imposes only 0.031 and 0.034% total power overhead for the probabilistic and the intolerant applications respectively. The total area overhead is 0.022%.

## 3.2 PVT Variations

In this section, we analyze the delay variations caused by PVT variations on the paths of the 32-bit in-order LEON3 processor compliant with the SPARC V8 architecture. This choice is keeping in view of the recent trends toward array processor architectures containing many simple RISC cores, e.g., GPUs [7], TILERA [8], and Platform 2012 [9, 10]. More importantly, the availability of an advanced open-source RISC core with full back-end details is critical to accurate variation analysis. We note that other efforts for complex high-performance cores such as IBM POWER6 also con-

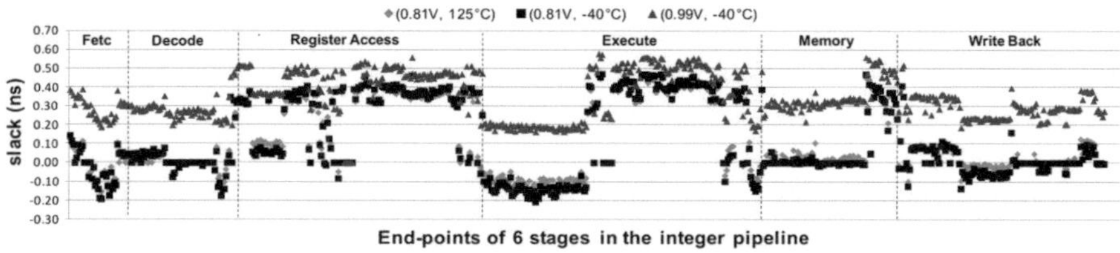


**Fig. 3.1** Nonuniform slack variation of the integer pipeline stages caused by PVT with cycle time at 0.83 ns

firm that vulnerability is not uniform across the instructions set [11]. While different instruction sets will lead to different grouping of instructions depending upon the processor architecture and implementation, our methodology can be applied as long as there is a nonuniform vulnerability across the instructions.

Specifically, the effects of a full range of dynamic variations (an industrial temperature range of −40–125 °C, and a voltage range of 0.72–0.99 V) as well as static process parameters variations (die-to-die and within-die) are analyzed on all paths throughout the entire integer pipeline of LEON3. Figure 3.1 illustrates the delay variation in the six stages of the pipeline that results in positive/negative slacks for the flip-flops connected to the endpoints of the paths. The cycle time is set at 0.83 ns to meet the timing requirement of the typical-corner (0.9 V, 25 °C, TT). A higher voltage of 0.99 V results in shorter delay (positive slack), while the lower temperature leads to a higher delay in the low-voltage region below 0.9 V, since MOSFET drain current decreases when the temperature is decreased in nanometer CMOS technologies [12]. In addition to these dynamic operating conditions, the static process variations exacerbate the delay variation across various pipeline stages: Sect. 3.2.2 describes the details of modeling the process variations. Given such variations across operating conditions and across different parts of the design, an adaptive guardbanding of the operating frequency is useful to ensure the error-free operation. Such a guardband can be much less conservative than a statically determined guardband. We divide pipeline paths into two groups: (a) Vulnerable Paths (VP): A set of paths that always require correct timing and any delay variability may result in catastrophic architectural failures and consequently visible errors in the outputs of a program; and (b) Invulnerable Paths (IP): A set of paths that do not require 100% timing correctness. The delay variation in IP does not cause catastrophic architectural failures since it affects only the vector of elastic outputs. The vector of elastic outputs does not require the complete numerical correctness. Thus, the delay variation in IP may degrade of the quality of fidelity metrics of the probabilistic applications. Specifically for LEON3 pipeline shown in Fig. 3.1, a 20% voltage variation results in many negative slack values at the endpoints of the fetch and decode stages which causes the wrong instructions to be executed. Thus, the paths that lie in these stages are considered as VP and must always meet the setup time of flip-flops in PVT variation. On the other hand, the scenario for IP is different. For example in the execution stage, some endpoints do not suffer from delay variation at all (those paths with a positive slack), and some endpoints have negative slack when voltage variation occurs. The execution stage has much more flexibility to deal with delay variation as long as it can produce an acceptable fidelity metric.

In Sect. 3.3.1, we present guardbanding technique that seeks to guardband VP for error-free operation, and at the same time effectively reduces the cost of guardbands on IP against fidelity metric of programs that are tolerant to imprecise and approximate computations. The tolerance levels can be specified based on algorithmic classifications such as RMS [13]. Section 3.4 also covers another adaptive guardbanding technique for intolerant applications in the general case.

### 3.2.1 Conventional Static Timing Analysis

Conventional static timing analysis (STA) calculates the maximum delay variation using the worst-case corner, by simply combining the absolute worst-case combination of the process, voltage, and temperature parameters. The cycle time is finely varied to observe the behavior of the pipeline stages. The number of failed paths (i.e., paths with negative slack) for each stage using the STA in the worst-corner (0.72 V, 0 °C, Slow NMOS-Slow PMOS) is shown in Fig. 3.2. Increasing the cycle time from 1.8 to 2.25 ns reduces the number of failed path from hundreds of thousand paths to zero path for all stages except the execution stage which has a higher delay. The execution stage needs 10% more guardbanding, i.e., the clock cycle of 2.5 ns. Further, our earlier analysis [14] shows that the execution and memory stages are highly vulnerable to dynamic variations. By setting the cycle time at 2.25 ns, we guarantee that no path will fail within the fetch, decode, register access, memory, and write back stages even in the worst-case process parameter variation. The paths in these stages are considered as VP because: (i) any failure in fetch or decode stages

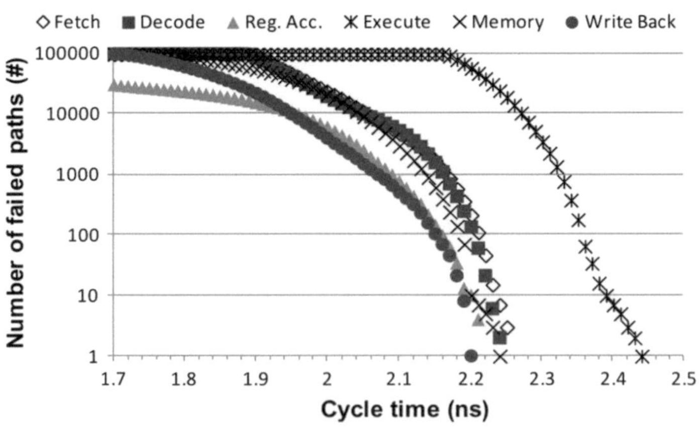


**Fig. 3.2** Number of failed paths of LEON3 pipeline using STA

may cause the wrong instructions to be executed that cannot be masked even within the probabilistic application; and (ii) any failure in the register/memory/write back stages may cause an illegal access/operation on the memory/registers. It is therefore not surprising that both Intel resilient processor [15] and relaxed-reliability cores in ERSA [16] consider sufficient guardbanding in register stage, memory management unit, and L1 instruction cache. By sufficient guardbanding on VP through STA, the error-free operation of VP is guaranteed even if these paths display the worst-case process characteristics.

Unlike the above-mentioned stages, with the cycle time of 2.25 ns, the execution stage has few failed paths in the worst-case process variation. If these paths are activated through the pipeline, there is no guarantee for 100% timing correctness of the execution stage. This lack of timing correctness causes inaccuracies in the result of execution of some instructions, which can be masked by the error resilience at the application-level of the probabilistic applications [13], or proper software-based instruction duplication technique. Thus, these paths are considered as IP, since their violation might cause only application-level derating which strictly depends to the type of applications. In Sect. 3.3, we examine the likelihood of these violations, and the type of applications that can accept or refuse this kind of inaccuracies.

To observe the behavior of VP and IP on other architectures, we also consider a programmable graphic processing unit (GPU), THEIA [17]. THEIA features a multi-core architecture, and uses a ray casting approach for rendering. Every core in THEIA runs a local copy of the shader code, and has a pipelined SIMD unit, capable of performing fixed-point arithmetic on 3D vectors. Each core includes instruction entry point, fetch, decode, execute, and memory stages in conjunction with a control unit. Similar to Fig. 3.2, the number of failed paths for each stage of a THEIA's core is shown in Fig. 3.3. As shown, VP display no failure with a clock cycle of 3.2 ns, while the execution stage faces high number of failed paths. In fact, the execution

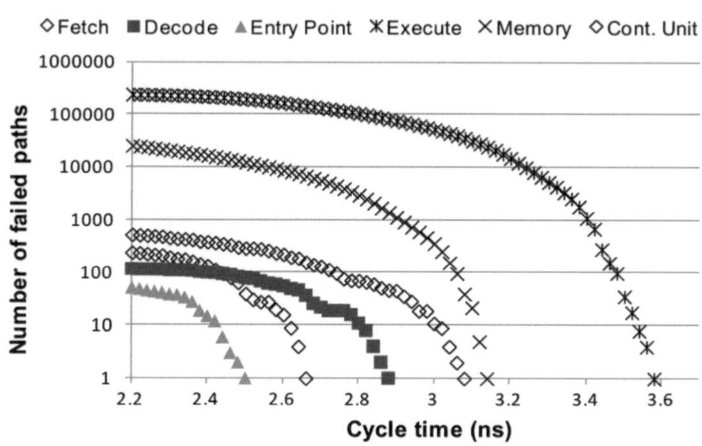


**Fig. 3.3** Number of failed paths of THEIA pipeline using STA

stage needs 14% more guardbanding compared to other stages. In comparison to LEON3, the execution stage of a THE-IA's core imposes higher guardbanding, since it performs vector fixed-point operations which involve more complex units than the scalar integer operation in LEON3.

Indeed, several researches show that execution stage is critical not only for in-order or SIMD architectures, but also for various VLIW and out-of-order architectures [18–20]. For instance, despite the prior-art assumption that the register file defines the clock frequency of a clustered VLIW processor, the realistic physical layout experiments for an 8-issue-slot VLIW pipeline show that it is the execution stage and its bypass network that limits the clock speed [18]. Although a clock frequency speedup is achieved by partitioning a single cluster into two clusters (thus a shorter bypass network); in subsequent clustering there is a steady decrease of the bypass network delay, hence the delay of functional units is a deciding factor in clock frequency since it occupies up to 85% of the clock period in an 8-cluster VLIWs [18]. M. Ozawa et. al. [20] also propose a cascade ALU architecture for out-of-order processors, in which the critical path lies in the ALU. Similarly, the ALU delay also determines the cycle time of a low-power out-of-order design [19].

### 3.2.2 *Variation-Aware Statistical STA*

Unlike the traditional STA, variation-aware statistical static timing analysis (SSTA) takes into account the actual distribution of the physical parameters instead [21]. As a result, the calculated slack distributions accurately reflect the true results obtained in silicon resulting in less pessimism in the analysis. The variation-aware SSTA is suitable for IP analysis where the processor does not need 100% timing correctness in case of the worst process variation. Our results illustrate the value of variation-aware SSTA. Figure 3.4 distinguishes the data arrival time of the execution stage of LEON3 for two operands using the worst-case STA versus the variation-aware SSTA. The operating condition is set for (0.81 V, 125 °C), and the process parameter for STA is set for the Slow NMOS-Slow PMOS (SS), while this parameter for variation-aware SSTA varies based on the process parameter variations supported by state-of-the-art commercial tools.

To perform an accurate design time SSTA, we use the variation-aware timing analysis engine of Synopsys PrimeTime VX [21], leveraging characterized parameters of 45 nm variation-aware TSMC libraries [22] derived from first-level process parameters by principal component analysis (PCA). PCA is a mathematical procedure that simplifies a data set by transforming a number of correlated parameters into a smaller number of uncorrelated parameters. After parasitic extraction from the physical design data, the die-to-die (D2D) and with-in-die (WID) process parameter variations are injected as normal distributions with zero means and standard deviations of $\sigma_{D2D} = 5\%$ and $\sigma_{WID} = 6.4\%$ [23]. Therefore, we change the variation components and analyze the delay variations with a given set of accurate variability models from commercial libraries [22], which are certainly more accurate than

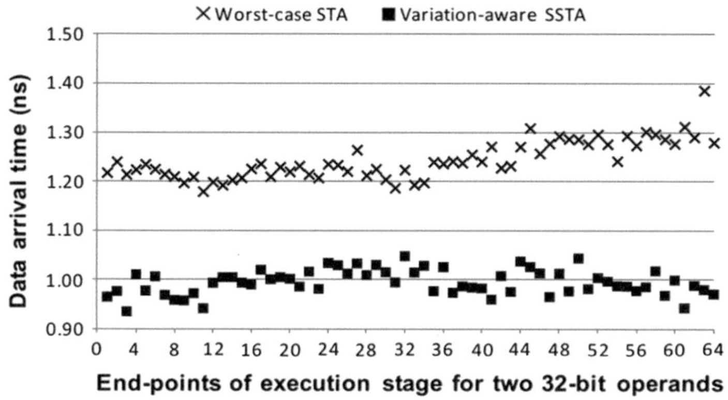


**Fig. 3.4** Variation-aware SSTA versus the worst-case STA

commonly used 'in-house model' extracted from predictive technology models [24]. As shown in Fig. 3.4, the data arrival time of the operands in the execution stage based on STA is upto 40% greater than the variation-aware SSTA due to pessimistic process parameters. For the fixed operating condition, STA results in 19% greater data arrival time on average compared to the variation-aware SSTA for the entire integer pipeline. These results set a baseline for the improvements from adaptive guardbanding techniques that raise the level of abstraction at which variability is addressed.

## 3.3 Error-Tolerant Applications

In moving from circuits to applications, we find a greater tolerance to failures simply because there is more contextual information available for recovery mechanisms to use. Given the increasing parallelism from hardware, the computer systems researchers have attempted to classify applications into core algorithmic categories such as RMS [13] that not only points to the structure of the computation but also a guidance on the degree of tolerance to individual data or even computational errors. While a comprehensive frame for classifying applications according to degree of data and control tolerance to error and variation is still an area of active research, adaptive guardbanding proposed here does bring us a step closer to tie the mitigation of PVT guardbands to the type of applications.

### 3.3.1 Analysis of Adaptive Guardbanding for Probabilistic Applications

For error-tolerant, or probabilistic, applications, the key idea is to guarantee the error-free operations of the paths that are vital for ensuring timing of the VP, while reducing the cost of guardbanding for the rest of the paths (IP). The adaptive guardbanding for the probabilistic applications dynamically decides on the cycle time based on the operating conditions, while guaranteeing the accuracy of the fidelity metric above a user-defined threshold ($U_T$) for the acceptable output. Timing error due to the delay variation in IP may alter the vector of elastic outputs ($O_E$). A fidelity metric of a probabilistic application P, FP (I, $O_E$) is associated with its input I and the corresponding $O_E$. The execution of application P with input instance I in the presence of delay variation is acceptable iff (A)∧(B)∧(C). The predicates (A)–(C) are defined as:

$$\begin{aligned} &(\mathrm{A})\ \mathrm{F}_P(I, O_E) \geq \mathrm{U_T} \\ &(\mathrm{B})\ \neg\exists\, path_j \in VP \mid \mathrm{Slack}_{\mathrm{STA}}(path_j) < 0 \\ &(\mathrm{C})\ \neg\exists\, path_k \in IP \mid \mathrm{Slack}_{\mathrm{SSTA}}(path_k) < 0 \end{aligned} \tag{3.1}$$

Specifically, the cycle time, for every operating condition is adjusted in such a way to satisfy that all paths in VP always meet the setup time of flip-flops even in the worst-case process parameter variation using STA (B); and that the paths in IP will not miss the setup time of any connected flip-flop, in a statistical sense, using the variation-aware SSTA (C). These two criteria guarantee the semantically correct execution of application P, e.g., an addition instruction is always executed as an addition instruction but it might generate inaccurate results, in case of large variations. To satisfy (A), the fidelity metric has to be greater than the UT, thus guarantees the acceptable accuracy from the applications' point of view. For a given application P, the application writer is responsible to tune the acceptable threshold based on the end user's requirements [25].

The adaptive guardbanding dynamically sets the cycle time to meet (A)–(C) requirements to mitigate the inter-corner variations for a given operating condition. The assigned cycle time guarantees the error-free operation of VP even in the worst-case process parameters variation, certified by STA. However, the guardband provided by the adapted cycle time cannot guarantee 100% timing correctness of IP within the execution stage in case of absolute worst-case combination of process parameters. This might cause inaccuracy in the result of the executed instruction. If the executed instruction produces $O_E$ (thus affecting the fidelity metric), the predicate (A) guarantees that the program can produce an acceptable fidelity metric. On the other hand, if the executed instruction is a critical instruction, the proper application-level correctness techniques [25] is applied to identify the critical control flow instructions. The critical instructions are statically duplicated during compile time which guarantees the error-free execution in a fixed operating condition.

**Table 3.1** Effectiveness of adaptive guardbanding for the probabilistic applications under dynamic variations

| Volt. (V) | Temp. (°C) | Cycle time (ns) | The worst slack of *execution* (ns) | | | |
|---|---|---|---|---|---|---|
| | | | mean | std-dev | p99 | p01 |
| 0.99 | −40 | 0.80 | 0.247 | 0.028 | 0.325 | 0.196 |
| 0.81 | −40 | 1.35 | 0.400 | 0.057 | 0.565 | 0.302 |
| 0.81 | 125 | 1.32 | 0.451 | 0.076 | 0.638 | 0.281 |
| ... | ... | ... | ... | ... | ... | ... |

We use SSTA methodology to analyze the effect of within-die and die-to-die process parameters variations. It dynamically sets the cycle time depends to the operating conditions as shown in Table 3.1. For example, as soon as detecting the operating condition at (0.99 V, −40 °C), the adaptive guardbanding decreases the cycle time from 2.5 ns, calculated by the worst-case STA for (0.81 V, 0 °C, SS), to 0.8 ns. This cycle time of 0.8 ns meets all timing requirements of VP, and at the same time provides positive slack for the execution stage in a statistical sense. As shown in the fourth column of Table 3.1, based on SSTA, the adaptive guardbanding strategy works well even with die-to-die and within-die process variation, while the paths are experiencing a full swing for voltage and temperature, and provides the positive slacks for the slowest path of the execution unit. Furthermore, the 1st percentile (p01) values are quite far from the zero slack, thus implying that the probability that actual slack of the path in the execution stage will be less than or equal to p01 value is 0.01.

The probability density functions of the slack value of top 1,000 critical paths within the execution stage are analyzed, at three operating conditions using the assigned cycle time in Table 3.1. All slack values are always positive when pipeline experiences a full swing in voltage ($\Delta V = 0.18$ V) and temperature ($\Delta° C = 165$ °C). If an IP path in the execution stage is faced with the worst-case combination of process parameters, and does not meet the timing requirement, the effects of such variations may manifest itself as an error in a bit of the output vector. Depending upon the positional significance, a probabilistic application may tolerate errors in low-order bits; for the high-order bits of the execution stage, there is little likelihood of having errors even in a full swing of the operating conditions, as the smallest p01 slack values are quite positive: 0.22 ns/0.37 ns at (0.99 V, −40 °C)/(0.81 V, 125 °C). The application writer can trade-off between the end user's accuracy requirements versus the cost of guardbanding using profiling and tuning mechanisms, thus satisfying predicate (A). The trade-off between the cycle time and the probability of having a failure in the execution paths is shown in Fig. 3.5. As shown, a higher cycle time results in lower probability of failure and thus a lower timing error rate. Therefore, the desired cycle time can be extracted to match with the tolerable error of the application. If the tolerable error of the application changes over different phases of the application, the policy of applying the adaptive guardbanding can be reprogrammed accordingly during the execution of the application.

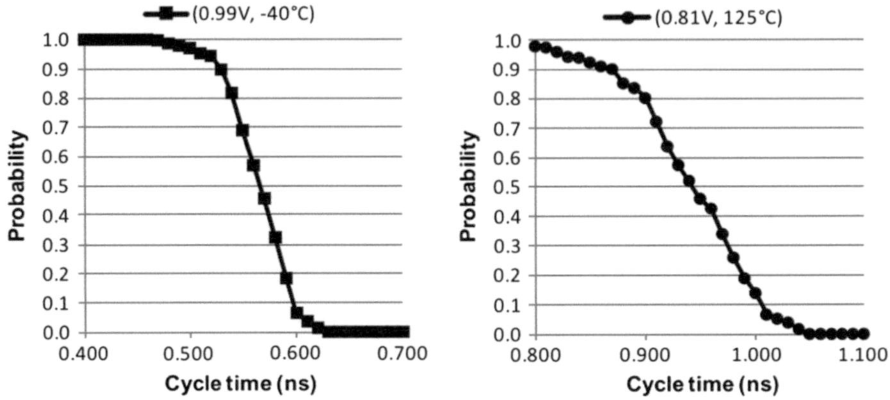


**Fig. 3.5** Trade-off between the cycle time and the probability of having a failure in the execution stage

## 3.4 Error-Intolerant Applications

### *3.4.1 Sequence-Level Vulnerability (SLV)*

Unlike the probabilistic applications, applications in general do not have such inherent algorithmic and cognitive tolerance thus even a single bit error in the execution unit could crash a program. We consider this class of applications as intolerant applications that require complete numerical correctness. Intolerant applications cover most of the general purpose applications, and even those probabilistic applications that there is no domain expert to define and analyze their fidelity metrics parameters. Therefore, the adaptive guardbanding for the intolerant applications has to guarantee 100% timing correctness for VP as well as IP. To alleviate such expensive constraint imposed by the intolerant programs, we have earlier defined the notion of instruction-level vulnerability (ILV [14]) to dynamic voltage and temperature variations in order to expose and use variation in architectural/compiler optimizations. Equation 3.2 defines ILV as a function of current operating voltage and temperature (V,T), and the corresponding class of an instruction ($\text{inst}_i$) determined by partial function of $\phi$. ILV is computed as the number of cycles with a failed path over the total Monte Carlo simulated cycles for the $\text{inst}_i$ in [14].

$$ILV = \Im(\phi(\text{inst}_i), V, T)$$

$$\phi(\text{inst}_i) = \begin{cases} ClassI & \text{if inst}_i \in \text{ALU instructions} \\ ClassII & \text{if inst}_i \in \text{MEM instructions} \\ ClassIII & \text{if inst}_i \in \text{HW MUL/DIV instructions} \end{cases} \tag{3.2}$$

In fact, ILV data in [14] partitions integer SPARC V8 ISA (except control instructions) into three classes: ClassI consists of ALU instructions; ClassII covers all memory (MEM) instructions; and ClassIII has hardware multiply/divide (MUL/DIV) instructions. As shown in Eq. 3.3, ILV indicates that the classes of instructions have different levels of vulnerability to dynamic variations depending on the way they exercise the nonuniform critical paths across the various pipeline stages. For instance, the hardware MUL/DIV instructions have a higher vulnerability in comparison to MEM instructions.

$$\begin{aligned}&\forall(V,T):\\&ILV(ClassI,V,T)\leq ILV(ClassII,V,T)\leq ILV(ClassIII,V,T)\end{aligned} \tag{3.3}$$

ILV does not cover the control instructions, because the characterization of a control instruction itself is meaningless unless it is considered within a sequence of instructions that affect the control instruction. Hence, we extend the notion of ILV; we introduce the notion of sequence-level vulnerability (SLV) to expose dynamic variation in Eq. 3.4. Different sequences of instructions exercise the critical paths of the pipeline differently resulting in various levels of vulnerability. The vulnerability of a sequence of instructions ($\text{seq}_i$) varies based on the class of instructions that it contains. SLV is also a function of current operating voltage and temperature to capture inter-corner dynamic variations. Therefore, SLV reflects the manifestation of variability-induced timing errors in the specific software context which is a sequence of instructions.

$$SLV=\Im(\varphi(\text{seq}_i),V,T) \tag{3.4}$$

### 3.4.2 SLV Characterization

To avoid an exponentially growing number of sequences for evaluations of SLV, the highly frequent sequences are extracted from various type of applications. We have profiled a large set of general purpose benchmarks containing 32 different applications, include MiBench [2], Parsec [6], Scimark2 [26], MediaBench [3], and CoreMark [5] benchmarks. The binaries of applications were dynamically instrumented. This allows us to extract the highly frequent sequences of the instrumented instructions as well as their operands distribution for the memory, and ALU instructions. This operands distribution helps to create the realistic values for the operands of the instructions. To distinguish sequences, a window of three instructions is considered since there are three stages before reaching the execution stage of LEON3. Then, for the sake of illustration, the top 20 highly frequent sequences are considered for the SLV analysis that are shown in Table 3.2.[1] After the sequence extraction, a sequence generator applied Monte Carlo method for each of top 20 sequences, utilizing the operands distribution instrumented from the aforementioned benchmarks. Therefore,

[1] We later show our method is not limited to the top sequences and a sequence with a length of three instructions (L = 3).

**Table 3.2** Extracted highly frequent sequences of instructions

| Seq. # | 1 | 2 | 3 | 4 | 5 | 6 | 7 | 8 | 9 | 10 |
|---|---|---|---|---|---|---|---|---|---|---|
| $Inst._i$ | ld | st | ld | ld | ld | st | ld | call | ld | call |
| $Inst._{i+1}$ | ld | st | bz | bz | st | ld | ld | st | ld | st |
| $Inst._{i+2}$ | ld | st | sub | and | ld | st | bz | st | sub | ld |
| Seq. # | 11 | 12 | 13 | 14 | 15 | 16 | 17 | 18 | 19 | 20 |
| $Inst._i$ | bz | ld | sub | and | sub | and | and | sub | sub | ALU |
| $Inst._{i+1}$ | st | and | bz | bz | add | add | sub | sub | and | ALU |
| $Inst._{i+2}$ | ld | st | bnz | bnz | bz | bz | bz | bz | bz | ALU |

large samples of highly frequent sequences for SPARC ISA have been generated, including ALU, MEM, and control instructions.[2]

Then, to accurately evaluate SLV under different operating conditions, these sequences were fed to the post-layout simulations where the delay of the layout implementation of the processor is back-annotated. Therefore, SLV is calculated for every individual sequence under a full range of operating conditions and cycle times to enable use of dynamic variations on sequences of instructions. To evaluate SLV, $seq_i$ is run through the pipeline while varying the operands of the instructions using the following algorithm:

For $seq_i \in$ list of high-frequent sequences
  For $(V, T) \in \{(0.72\,V, -40\,°C), ..., (0.99\,V, 125\,°C)\}$
    For $Cycle\_Time \in \{1.0\text{ ns}, ..., 3.0\text{ ns}\}$
      For *operands* $\in$ list of operands
        Compute $SLV\ (seq_i, V, T, Cycle\_Time)$

The SLV for each $seq_i$ at the operating condition (V, T) with cycle time is quantified in Eq. 3.5, where $N_i$ is the total number of clock cycles in Monte Carlo simulation of $seq_i$ with random operands; and $Violation_j$ indicates if there is a violated stage at clock $cycle_j$ or not. In other terms, SLV is defined as the total number of violated cycles over the total simulated cycles for the $seq_i$. If any of the six stages have one or more violated flip-flop at clock $cycle_j$, we consider that stage as a violated stage at $cycle_j$ since there is at least one activated critical path for $seq_i$ at $cycle_j$ that is slow enough to miss the setup time of a flip-flop. Intuitively, if $seq_i$ runs without any violated path, SLV is zero; on the other hand, SLV is one if for every cycle $seq_i$ faces at least one violated path in any stage.

[2]The rest of ISA needs the floating-point and coprocessor units which are not available neither in our core nor in [15].

$$SLV(\text{seq}_i, V, T, Cycle\_Time) = \frac{1}{\text{N}_i}\sum_{j=1}^{\text{N}_i} \text{Violation}_j$$

$$\text{Violation}_j = \begin{cases} 1 & \text{If any stage violates at cycle}_j \\ 0 & \text{otherwise} \end{cases} \tag{3.5}$$

Figure 3.6 shows the SLV values of the top sequences under a wide range of voltage and temperature variations while the cycle time is finely varied (steps of 10 ps). The SLV values are 0 during the long cycle times, as the cycle time decreases the SLV values increase toward 1 because the sequences experience higher timing violations. Let us first examine the behavior of the sequences under the full range of temperature variation (Fig. 3.6b and c). At the temperature of 125 °C, all sequences have a SLV of 0 with clock cycle 1.35 ns. By decreasing the cycle time beyond 1.33 ns, $\text{seq}_1$–$\text{seq}_{19}$ start to incur the timing violation as their SLV values increase, while $\text{seq}_{20}$ is displaying a SLV of 0 until decreasing the cycle time to 1.28 ns. This trend also persists under $\Delta\text{T} = 165\,°\text{C}$ temperature fluctuation with a shift in cycle time (Fig. 3.6c). As shown, these sequences are partitioned into two classes based on the SLV values. The $\text{seq}_1$–$\text{seq}_{19}$ have higher within-corner SLV values, while the $\text{seq}_{20}$ has lower within-corner SLV values.

Let us now examine the SLV values under dynamic voltage variations (Fig. 3.6a and b). A similar pattern of within-corner SLV variations is observed: the $\text{seq}_1$–$\text{seq}_{19}$ show higher SLV values compared to the $\text{seq}_{20}$ at equal cycle times. This classifies the $\text{seq}_1$–$\text{seq}_{20}$ into two classes of sequences: ClassI and ClassII. As defined in Eq. 3.6, ClassI is a sequence of instructions of length L in which every instruction has an ILV class of ClassI. In other words, when a sequence of instructions is composed of only ALU instructions, the sequence is classified as ClassI; otherwise it is classified as ClassII. Therefore, an instruction within the sequence of ClassII can be any instruction, including MEM, MUL/DIV, and various control instructions. For every operation condition (V, T), ClassI has a lower SLV (thus needs lower guardband) in comparison to ClassII.

$$\forall(V, T, \text{seq}_i) : SLV(ClassI, V, T) \leq SLV(ClassII, V, T), \text{ s.t.}$$

$$\varphi(\text{seq}_i) = \begin{cases} ClassI & \forall \text{inst}_j \in \text{seq}_i | \phi(\text{inst}_j) = ClassI, 2 \leq j \leq L \\ ClassII & \text{otherwise} \end{cases} \tag{3.6}$$

Based on our analysis for the highly frequent sequences, as shown in Fig. 3.6, the $\text{seq}_{20}$ is classified as ClassI, while the $\text{seq}_1$–$\text{seq}_{19}$ are among ClassII. The $\text{seq}_{20}$ has a lower SLV compared to all sequences in ClassII; since its instructions do not involve the critical paths of the memory and control (integer code conditions) components. Thus, we see that the SLV value of the two classes of the sequences at the same corner and with the same cycle time is not equal because their instructions do not uniformly exercise the various critical paths of the pipeline. We know that the vulnerability of instructions is not uniform [14]. Sequences in ClassII need higher guardbands

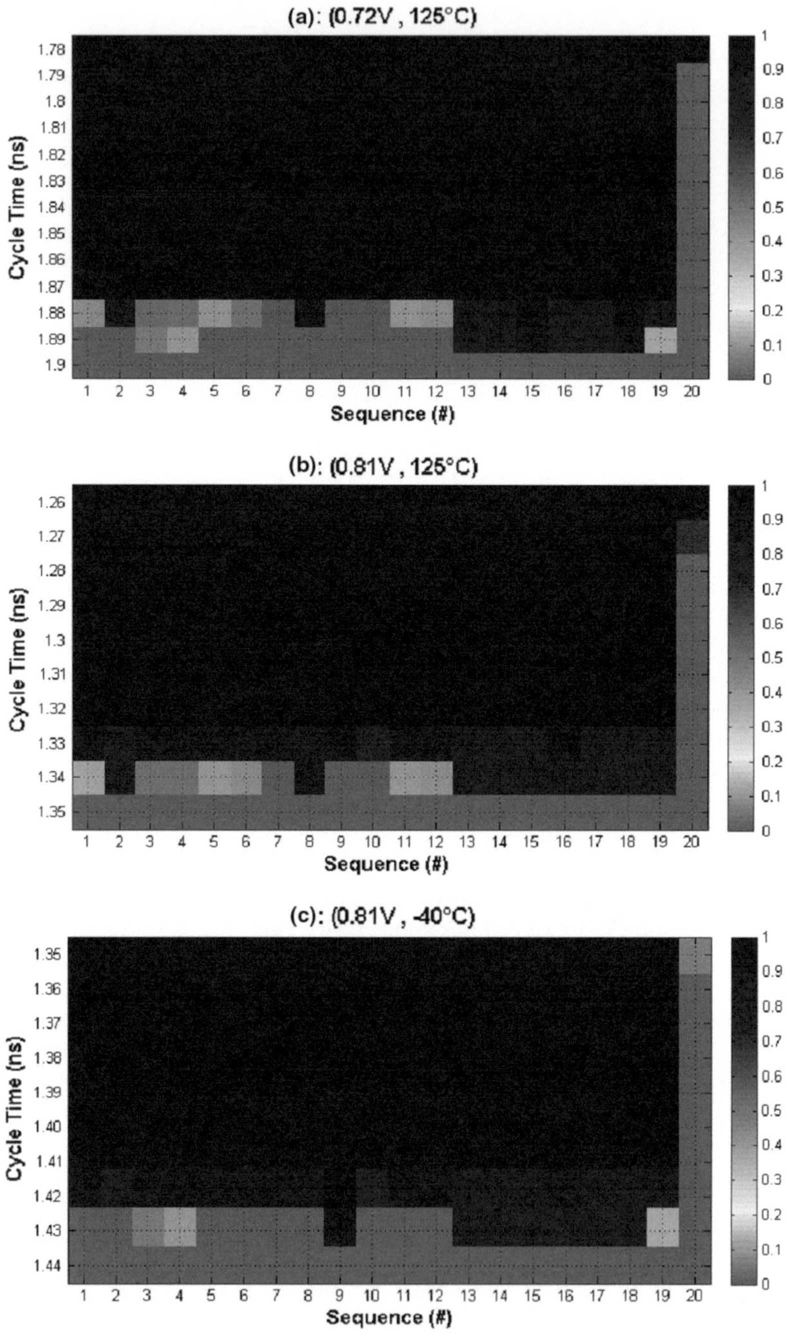


**Fig. 3.6** Intra-corner SLV to dynamic variations ($\Delta$T = 165 °C and V = 0.09 V); **a** (0.72 V, 125 °C), **b** (0.81 V, 125 °C), **c** (0.81 V, −40 °C)

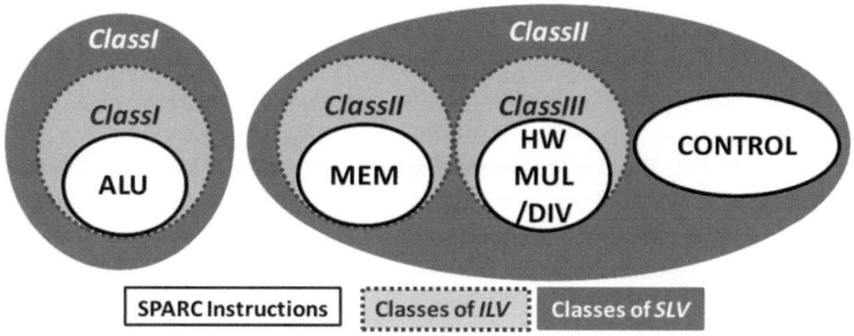


**Fig. 3.7** ILV and SLV classification for integer SPARC V8 ISA

in comparison with ClassI, mainly because in addition of ALU's critical paths, the critical paths of memory are also activated for the load/store instructions as well as the critical paths of integer code conditions for the control instructions. As a result, in the same corner, sequences in ClassI run faster, thanks to their all ALU instructions which only exercise critical paths of the ALU component.[3] Figure 3.7 summarizes ILV and SLV classification.

This intra-corner SLV enables the adaptive guardbanding to set the cycle time for each class of sequences accordingly, and thus eliminate the conservative guardbands across sequences up to 6%. Therefore, for intolerant applications, the adaptive guardbanding adjusts the cycle time depending upon the classes of the sequence, and the current operating conditions to make sure that the processor runs at the fastest speed compatible with both current hardware and software conditions. We classify any noncharacterized sequence out of the analyzed highly frequent sequences as ClassII, thus it will have appropriate timing guardband in case of activation of the critical paths of non-ALU components. Relaxing the guardband can also be applied to any sequence of ClassI with a length of two ALU instructions ($\text{ClassI}_{L=2}$) or more (N) ALU instructions stream ($\text{ClassI}_{L=N}$). These chains of ALU instructions exercise the critical paths within only ALU component, therefore, for a given operating condition as shown in Eq. 3.6, the SLV values of $\text{ClassI}_L$ for $L \in \{2,3,\ldots,N\}$ are equal. This classifies ALU sequences into the same class of the sequences with consistency across a wide range of corners.

## 3.5 Adaptive Guardbanding

We propose a guardbanding technique that dynamically decides on the cycle time based on the Application's Type, the Instruction Sequence, and the operating conditions (V,T), to maximize performance. To ensure necessary observability, our approach employs on-chip low-overhead operating condition monitors using

[3] ALU does not include the hardware multiply and divide units.

CPM [27]. POWER7 results show that five CPMs per each core are sufficient to finely capture PVT variation [28]. For controllability, a fast adaptive clocking circuit consisting of three phase-locked loops (PLLs) is leveraged. Each PLL is running at independent frequencies, and a multiplexer quickly switches between them in a single cycle [29, 30]; therefore ultra-fast frequency changes are possible and PLL lock time is not an issue. This is well suited to mitigate the inter-corner dynamic variations where the timing guardbanding across corners are far apart. To mitigate the intra-corner guardband between the two classes of sequences, a finer clock speed adaptation is required which can be supported by an all-digital PLL. For instance, [30] proposes an all-digital PLL that provides multiple equally spaced clock phases with a small tuning step size of a few picoseconds; these phases are switched in a glitch-free reverse switching scheme. A phase switching frequency division architecture is also used to generate subinteger division ratios and thus a larger variety of output frequencies [35]. These circuits techniques support very fast adaptation of the clock speed of the processor in immediate response to changes in the operating corners, various sequences of instructions, and the type of applications. The adaptive guardbanding adjusts the cycle time as defined in Eq. 3.7.

$$Cycle\_Time = \Im(Application's\, Type,\ Instruction\, Sequence,\ V,\ T) \tag{3.7}$$

where Application's Type is probabilistic or intolerant; Instruction Sequence is the type of sequence which is either ClassI or ClassII; V and T are discretized current operating conditions reported by on-chip CPM sensors; the function is represented by a programmable lookup table (PLUT) as shown in Table 3.3. The PLUT is a fully combinational module in the pipeline.[4] It is programmable through the memory-mapped I/O in arbitrary epochs of the post-silicon stages. The PLUT is connected to CPM (for monitoring the current operating condition (V, T)), the fetch stage (for monitoring the Instruction Sequence), and the single-cycle adaptive clocking module (for setting the cycle time). The Application's Type is also set at the start of running the application via memory-mapped I/O. The adaptive guardbanding monitors these four parameters every cycle, and then sends corresponding commands to the clock speed adjustment circuit to make sure that processor always runs at the fastest speed compatible with these conditions.

As shown in Table 3.3, there is no intra-corner cycle time adaptation for the probabilistic application. The within-corner correct execution is guaranteed by static duplication of the critical instructions which is the application-aware version of the multiple-issue instruction replay [15]. Therefore, for the probabilistic application we do not require an online hardware recovery unit, and avoid the frequent changing of the cycle time within an operating corner.

In our experiments, for characterization of the PLUT, we have used six sign-off operating corners available on an advanced real-life technology library [22]. PLUT conservatively matches a surrounding operating condition if the discretized reported operating condition does not appear in the PLUT. Note, this is conservative for few

---

[4]Note that PLU can be characterized and then optimized during design time stage depending upon the range of operating conditions and application's type.

**Table 3.3** PLUT for adaptive guardbanding

| Application's type | Instruction sequence | Voltage (V) | Temperature (°C) | Cycle time (ns) |
|---|---|---|---|---|
| Probabilistic | – | 0.99 | 0 | 0.78 |
| Probabilistic | – | 0.81 | 125 | 1.32 |
| Probabilistic | – | 0.72 | 125 | 1.55 |
| Intolerant | *ClassII* | 0.81 | −40 | 1.44 |
| Intolerant | *ClassI* | 0.81 | −40 | 1.36 |
| Intolerant | *ClassI* | 0.72 | 125 | 1.80 |
| ... | ... | ... | ... | ... |

points in the PLUT, but will converge to ideal, while still being safe, if semiconductor fabrication process provides more characterized operating corners. Furthermore, for the intolerant applications, the adaptive guardbanding considers the worst-case process variation, and considers a conservative guardband (as safe as ClassII) on the noncharacterized sequence of instructions (sequences out of $seq_1$–$seq_{20}$), thus guarantees 100% numerical correctness for the intolerant applications. As shown in Table 3.3, the PLUT assigns different cycle times to various types of applications at the same operating condition. Inherent resiliency of the probabilistic applications indicates that these can tolerate inaccuracies, while the intolerant applications do not accept such inaccuracies. Therefore, when running an intolerant application the sufficient guardbanding is guaranteed for IP as well.

## 3.6 Experimental Results

The experimental methodology for STA, and the variation-aware SSTA are described using Fig. 3.8 that shows both design time and runtime flows. During the design time analysis, the open-source synthesizable VHDL code of LEON3 [31] and Verilog description of the PLUT module have been synthesized with the TSMC 45 nm technology library, the general purpose process. The synthesized core enables the variation analysis of paths of the integer parallel pipeline unit, as well as the L1 instruction cache (I$) and the L1 data cache (D$), unlike the resilient core [15] that only considers the integer unit. The front-end flow with normal VTH cells has been performed using Synopsys Design Compiler with the topographical features enabled, while Synopsys IC Compiler has been used for the back-end. The design is optimized for performance with the tight timing constraints, e.g., the clock period of 1.2 ns. For SSTA, the sign-off stage has been made with variation-aware timing analysis of Synopsys PrimeTime VX, leveraging characterized parameters of TSMC 45 nm variation-aware libraries discussed. The dynamic variations are also analyzed utilizing the six accessible TSMC characterized sign-off corners. Finally, for the post-layout simulations Mentor Graphics ModelSim is employed.

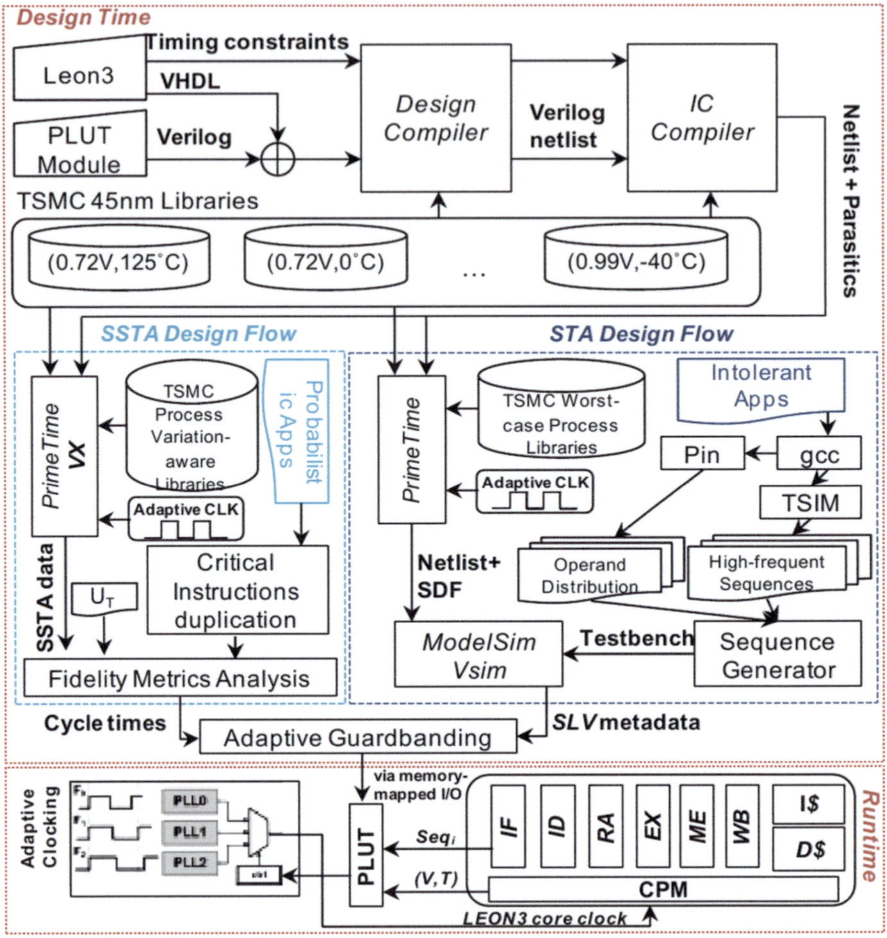


**Fig. 3.8** Methodology for the adaptive guardbanding

At the runtime, in every cycle, the PLUT module sends the desired cycle time to the adaptive clocking circuit utilizing the characterized SLV of the current sequence and the operating condition monitored by CPM. For detecting the current sequence, the PLUT looks at a window of three instructions (available on IF, ID, RA stages), thus it detects the class of the current instructions sequence before they reach the execution stage (the stage that needs more guardbanding as shown in Fig. 3.2. The previous stages (IF, ID, RA) are in a safe guardband, thus they will not have any failure if a sequence of ClassI/ClassII is running while the cycle time is set for a ClassII/ClassI. If the pipeline architecture does not have enough stages before the execution, the prefetch buffer [32] can be monitored instead. By detecting changes in the class of sequences, the single-cycle adaptive clocking circuit sets the core frequency accordingly. If an adaptive clocking circuit has long-latency clock switching, the PLUT can look ahead of a prefetch buffer coupled with phase prediction techniques to be able to decide about the desired core frequency in advance. Note that the core consists of the integer pipeline, L1 I$, and L1 D$ that are clocked by a single clock domain. Communication with L2 caches and uncore part can be done via globally asynchronous, locally synchronous interconnection supporting synchronization across multiple clock domains [9].

### 3.6.1 *Effectiveness of Adaptive Guardbanding*

Here, we investigate the effectiveness of our adaptive guardbanding technique when executing real-world applications.[5]

#### 3.6.1.1 Error-Tolerant Applications

As error-tolerant probabilistic applications, we have selected multimedia benchmarks from MiBench and MediaBench suites: H264 is a video decoder while Libmad is a MP3 decoder; Susan is an image recognition program; DCT, Huffman coding and Ycc2rgb are important kernels in the JPEG decoder; GSM implements a decoder for the GSM communications standard, and LDPC is a linear error correcting code. The appropriate fidelity metric analysis and application-level correctness technique based on [25] are performed to identify the critical control flow instructions of these applications. Then, the critical instructions are statically duplicated during compile time. Finally, the adaptive guardbanding determines the cycle time based on the given error probability 0.01% which can guarantee the acceptable fidelity metrics [25].

In the traditional worst-case design, the maximum throughput of applications is limited by 400 MIPS (million instructions per second), analyzed by the worst-case STA. Figure 3.9 shows the normalized throughput of the applications in various operating conditions, covering $\Delta V = 0.09\,V$ dynamic voltage variation and $\Delta T = 125\,°C$ temperature variation. In comparison with the worst-case design, the adaptive guardbanding changes the throughput of these applications from 0.95× to 1.9× depends to the current operating condition. Throughput of Rician is increased up to 1.9× at (0.81 V, 125 °C). On the other hand, throughput of Huffman coding at the operating condition of (0.72 V, 125 °C) is degraded by 0.95× because 69% of its instructions are the critical control flow instructions which are duplicated, and cancel out the benefit of faster execution of the total instructions. On average, the throughput of these applications is enhanced by 1.52×. This shows that utilizing SSTA and adapting to the operating conditions highly surpasses the traditional worst-case STA, and hides the overhead of the critical instructions duplication.

[5]For those applications that have encoder and decoder parts, we consider their back-to-back executions.

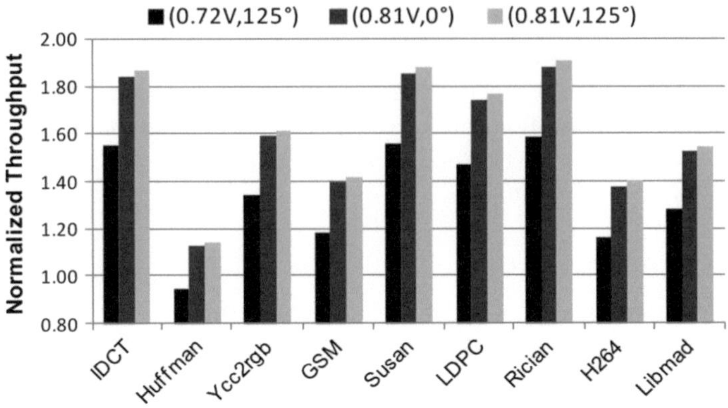


**Fig. 3.9** Normalized throughput improvement by utilizing SSTA compared to the worst-case design for probabilistic applications

#### 3.6.1.2 Intolerant Applications

For the intolerant applications, we have selected applications from six categories of MiBench, each suite targeting a specific area of the embedded market, including automotive, consumer devices, office automation, networking, security, and telecommunications. In addition, we have also considered EEMBC AutoBench [4] suite of benchmarks, suitable for embedded processor in automotive, industrial, and general purpose applications. Without loss of generality, every probabilistic application can be considered as an intolerant application and benefits from SLV utilization if there is no domain expert to define and analyze its fidelity metric. Figure 3.10 shows the percentage of sequences of ClassI with various lengths of ALU instructions, L$\in$ {2, 3,..., 7}, during execution of the intolerant applications. For instance, $\text{ClassI}_{L=2}$ shows the percentage of sequences that have exactly two consecutive ALU instructions, $\text{ClassI}_{L=3}$ represents sequences with just three consecutive ALU instructions, and so on. The compiler[6] optimizes the applications codes with `-O3` optimization option; and then the applications are profiled during execution using TSIM [33], a cycle-accurate instruction-level simulator. Figure 3.10.a shows on average 26% of the total executed sequences belong to ClassI, while the remaining sequences belong to ClassII. Patricia has the maximum number of sequences of ClassI, 35%. The adaptive guardbanding technique with the sequence detector of three instructions benefits from the sequences of ClassI with a length of 3 or more instructions.

Figure 3.10b shows the percentage of sequences of ClassI when the compiler utilizes loop unrolling technique. Loop unrolling is a loop transformation technique that attempts to increase speed of a program by reducing instructions that control the loop. It increases the number arithmetic instructions with regard to the memory and control flow instructions, at the expense of register pressure and program size. There-

[6]GNU Compiler Collection, version 3.4.4, with floating-point, mul/div emulation.

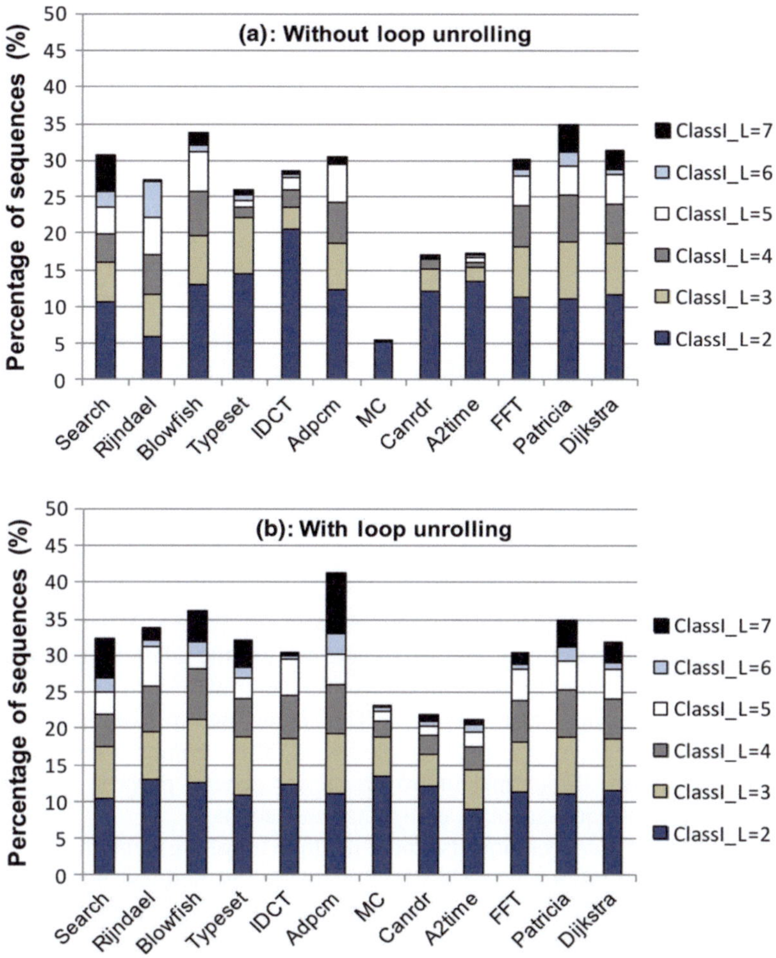


**Fig. 3.10** Percentage of sequences of ClassI during program execution: **a** without loop unrolling technique; **b** using loop unrolling technique

fore, applying the loop unrolling produces a longer chain of ALU instructions, and as a result the percentage of sequences of ClassI is increased up to 41% and on average 31%. Hence, the adaptive guardbanding benefits from this compiler transformation technique to further reduce the guardband for sequences of ClassI. Considering the sequence detection with a length of three instructions, the adaptive guardbanding reduces the cycle time for 20% of the executed sequences on average (up to 30% for Adpcm). Note that the adaptive guardbanding technique also reduces the guardband for the other sequences of ClassI with a longer length of three instructions, since each sequence of ClassI with L instructions is composed of two consecutive sequences with a length of L-1 instructions, considering the overlap between the two sequences.

**Table 3.4** Throughput improvement of the intolerant applications utilizing the adaptive guardbanding with loop unrolling

| Throughput improvement (×) | Only *SLV* (intra-corner) | | *SLV* + inter-corner | |
|---|---|---|---|---|
| | Max | Average | Max | Average |
| (0.72 V, 125°) | 1.04 | 1.03 | 1.36 | 1.35 |
| (0.81 V, 0°) | 1.06 | 1.05 | 1.80 | 1.78 |
| (0.81 V, 125°) | 1.05 | 1.05 | 1.88 | 1.87 |

Table 3.4 lists the maximum and the average throughput improvement of the adaptive guardbanding technique utilizing the loop unrolling during compilation phase of the intolerant applications. The throughput improvement is evaluated across various operating conditions. The second and the third columns of Table 3.4 show the maximum and the average throughput improvement of the applications utilizing SLV only within a fixed operating corner. Thus, the applications benefit from the higher rate of execution of the sequences of ClassI accomplished by the loop unrolling method. The last two columns show the maximum and the average normalized throughput (the worst-case design is the baseline) improvements utilizing SLV and inter-corner adaptation. In comparison with the worst-case design, the adaptive guardbanding enhances the throughput of these applications by a factor of 1.35× to 1.88× depending upon the current operating condition. This shows that utilizing the operating corner monitors and the online SLV coupled with offline compiler techniques can result in a significant throughput improvement for general purpose applications, where there is strict requirement on computational accuracy.

We compare our SLV technique (without the loop unrolling) with the code transformation technique proposed in [1] which pads the instructions sequence with a NOP instruction. The NOP padding eliminates the critical path activation for the forwarding paths of a processor for a read-after-write (RAW) register dependency. In other words, the result is no longer forwarded directly from the execution stage, it instead is forwarded a cycle later from the pipeline register in the memory stage. For comparison, we have identified the code sequences with a RAW register dependence and padded them with NOP instruction. Those NOP padded sequence are clocked as fast as the ClassI. The authors in [1] assume that they can clock that sequence 2.15× faster than the typical frequency of a processor, while Intel shows that in the resilient processor the clock can increase up to 0.16× in a fixed operating corner [15]; our results in Sect. 3.4.2 also indicates that intra-corner clock guardbanding for various sequences is bounded by 0.06×. Figure 3.11 shows the normalized (baseline is [1]) throughput of our adaptive guardbanding utilizing SLV by adapting the cycle for dynamic operating conditions and different classes of the sequences. On average, our technique achieves 1.65× higher throughput because [1] imposes one extra cycle for executing the NOP instruction, and does not adapt to the operating conditions. Figure 3.12 shows the energy overhead of the NOP padding across various operating

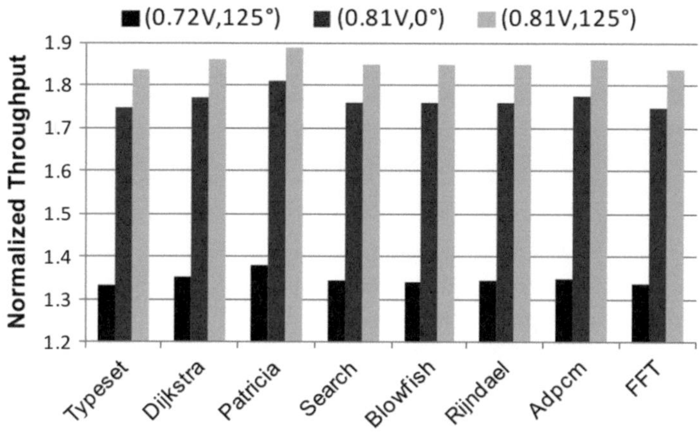


**Fig. 3.11** Normalized throughput improvement utilizing SLV compared to [1] for the intolerant applications

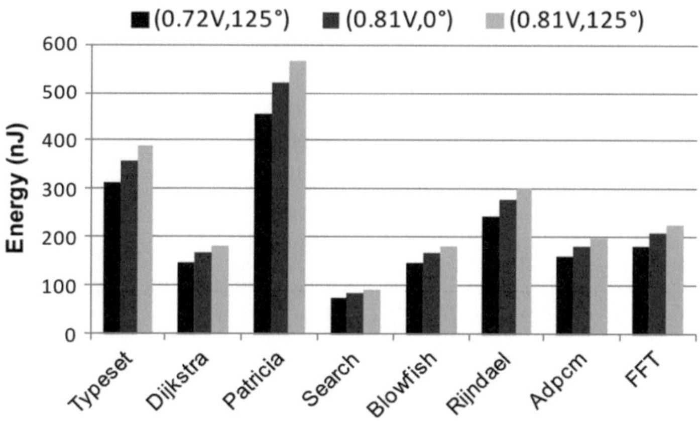


**Fig. 3.12** Energy overhead of NOP padding [1] across corners

corners. It imposes 74–564 nJ energy overhead, depending upon the number of NOP instructions and the current operating condition.

Multi-instruction code substitution, as another code transformation techniques in [1], is not applicable for an embedded RISC machine where there are almost no alternatives for representing an equivalent set of instructions, unless paying the expenses of intrusive pipeline modification, ISA extension, and leveraging co-processors. Nevertheless, there is a considerable performance and energy penalty for replacing a multi-instruction sequence with an equivalent set of instructions [34].

The common strategy in circuit techniques [15, 35] is to allow the timing errors to happen. Then, an extra cost is paid to compensate errors through the error recovery technique: the multiple-issue instruction replay imposes up to 28 extra recovery

**Table 3.5** Area and power overheads of adaptive guardbanding

| | LEON3 | Intolerant | Probabilistic |
|---|---|---|---|
| Total power (W) | 2.00E-01 | 6.79E-05 | 6.20E-05 |
| Leakage power (W) | 1.04E-02 | 1.24E-06 | 1.20E-06 |
| Total area (cell) | 744018 | 164 | 164 |

cycles per error [15]. This cost of recovery has shown to be high, thus leading to massive performance degradation if processor blindly relies on the error recovery in face of frequent timing errors, especially so in aggressive voltage overscaling and near-threshold computation. However, our proposed approach guarantees the correct execution at lower cost: (i) It proactively prevent timing errors on VP by applying the adaptive guardbanding across the operating corners and the sequence of instructions. For the error-intolerant applications, even if some residual timing error probability remains mainly because of using Monte Carlo method described in Sect. 3.4.1, our approach relies on the processor with error recovery capability that guarantees the correct execution with 100% numerical correctness. In this way, our online adaptive guardbanding implies that recovery actions will have to be undertaken in an extremely small number of cases, hence the recovery penalty is minimal. (ii) Our technique allows timing errors to happen on IP while meeting the application-specific requirements on computational accuracy for the error-tolerant applications, hence no penalty of recovery.

### *3.6.2 Overhead of Adaptive Guardbanding*

Table 3.5 lists the overhead of hardware implementation of the adaptive guardbanding technique. The area overhead in comparison to LEON3 core (including I$ and D$) is near-zero (0.022%). Five CPMs, as PVT sensors, occupy 0.12% area [28]. The adaptive guardbanding also imposes only 0.034%/0.031% average total power overhead for the intolerant/probabilistic applications, in the worst-case operating condition; the power leakage overhead is 0.012%. This coarse grained monitoring and adaptation approach is less intrusive and expensive and nicely complements the fine-grained approaches such as Razor and EDS.

## 3.7 Chapter Summary

A variation-aware cross-layer approach is presented that spans circuits, architectural pipeline to the applications. We propose a design time analysis in conjunction with the minimally intrusive runtime adaptive guardbanding technique to combat PVT variations while guaranteeing various applications demands on computation accuracy. We introduce the notion of sequence-level vulnerability (SLV) to capture variability characteristics that can be used by the compiler, runtime system, or even by the application programmer. The adaptive guardbanding technique enables an in-order RISC processor to run at the fastest speed compatible with the operating conditions, various sequences of instructions, and the type of applications. This increases the throughput of probabilistic applications upto 1.9× over the traditional worst-case design. Utilizing SLV achieves on an average 1.6× speedup for the intolerant applications, compared to [1], by adapting the cycle for dynamic variations and different instruction sequences. The concrete full layout results in TSMC 45 nm technology confirm that our technique incurs only 0.022, 0.012, and 0.034% overheads for the total area, leakage power, and total power respectively.

## References


1. G. Hoang, R.B. Findler, R. Joseph, Exploring circuit timing-aware language and compilation, in *Proceedings of the Sixteenth International Conference on Architectural Support for Programming Languages and Operating Systems, ASPLOS XVI* (ACM, New York, 2011) pp. 345–356
2. MiBench. http://www.eecs.umich.edu/mibench/
3. MediaBench. http://euler.slu.edu/~fritts/mediabench/
4. EEMBC benchmark consortium. http://www.eembc.org
5. CoreMark. http://www.coremark.org/home.php
6. PARSEC benchmark suite. http://parsec.cs.princeton.edu/
7. Whitepaper. NVIDIAS next generation CUDATM compute architecture: Kepler TM GK110 (2012)
8. S. Bell, B. Edwards, J. Amann, R. Conlin, K. Joyce, V. Leung, J. MacKay, M. Reif, L. Bao, J. Brown, M. Mattina, C.-C. Miao, C. Ramey, D. Wentzlaff, W. Anderson, E. Berger, N. Fairbanks, D. Khan, F. Montenegro, J. Stickney, J. Zook, Tile64 - processor: a 64-core SoC with mesh interconnect, in *IEEE International Solid-State Circuits Conference. ISSCC 2008. Digest of Technical Papers* (2008), pp. 88–598
9. D. Melpignano, L. Benini, E. Flamand, B. Jego, T. Lepley, G. Haugou, F. Clermidy, D. Dutoit, Platform 2012, a many-core computing accelerator for embedded SoCs: performance evaluation of visual analytics applications. In *49th ACM/EDAC/IEEE Design Automation Conference (DAC)* (2012), pp. 1137–1142
10. L. Benini, E. Flamand, D. Fuin, D. Melpignano, P2012: building an ecosystem for a scalable, modular and high-efficiency embedded computing accelerator, in *Design, Automation Test in Europe Conference Exhibition (DATE)* (2012), pp. 983–987
11. P.N. Sanda, J.W. Kellington, P. Kudva, R. Kalla, R.B. McBeth, J. Ackaret, R. Lockwood, J. Schumann, C.R. Jones, Soft-error resilience of the IBM POWER6 processor. IBM J. Res. Dev. **52**(3), 275–284 (2008)
12. R. Kumar, V. Kursun, Reversed temperature-dependent propagation delay characteristics in nanometer CMOS circuits. IEEE Trans. Circuits Syst. II: Express Briefs **53**(10), 1078–1082 (2006)
13. P. Dubey, Recognition, mining and synthesis moves computers to the era of tera. Technol. Intel Mag. **9**, 1–10 (2005)
14. A. Rahimi, L. Benini, R.K. Gupta, Analysis of instruction-level vulnerability to dynamic voltage and temperature variations, in *Design, Automation Test in Europe Conference Exhibition (DATE)* (2012), pp. 1102–1105

15. K.A. Bowman, J.W. Tschanz, S.L. Lu, P.A. Aseron, M.M. Khellah, A. Raychowdhury, B.M. Geuskens, C. Tokunaga, C.B. Wilkerson, T. Karnik, V.K. De, A 45 nm resilient microprocessor core for dynamic variation tolerance. IEEE J. Solid-State Circuits **46**(1), 194–208 (2011)
16. H. Cho, L. Leem, S. Mitra, ERSA: error resilient system architecture for probabilistic applications. IEEE Trans. Comput.-Aided Des. Integr. Circuits Syst. **31**(4), 546–558 (2012)
17. THEIA. http://opencores.org/project,theia_gpu
18. A. Terechko, M. Garg, H. Corporaal, Evaluation of speed and area of clustered VLIW processors, in *18th International Conference on VLSI Design* (2005), pp. 557–563
19. E. Gunadi, M. Lipasti, CRIB: consolidated rename, issue, and bypass, in *38th Annual International Symposium on Computer Architecture (ISCA)* (2011), pp. 23–32
20. M. Ozawa, M. Imai, Y. Ueno, H. Nakamura, T. Nanya, Performance evaluation of cascade ALU architecture for asynchronous super-scalar processors, in *ASYNC 2001. Seventh International Symposium on Asynchronus Circuits and Systems* (2001), pp. 162–172
21. PrimeTime VX user guide (2011)
22. TSMC 45 nm standard cell library release note, v 120a (2009)
23. S. Herbert, D. Marculescu, Characterizing chip-multiprocessor variability-tolerance, in *45th ACM/IEEE Design Automation Conference, DAC 2008* (2008), pp. 313–318
24. Predictive technology model (PTM). http://ptm.asu.edu/
25. J. Cong, K. Gururaj, Assuring application-level correctness against soft errors, in *IEEE/ACM International Conference on Computer-Aided Design (ICCAD)* (2011), pp. 150–157
26. SciMark v2.0. http://math.nist.gov/scimark2/
27. A. Drake, R. Senger, H. Deogun, G. Carpenter, S. Ghiasi, T. Nguyen, N. James, M. Floyd, V. Pokala, A distributed critical-path timing monitor for a 65 nm high-performance microprocessor, in *IEEE International Solid-State Circuits Conference. ISSCC 2007. Digest of Technical Papers* (2007), pp. 398–399
28. M. Floyd, M. Allen-Ware, K. Rajamani, B. Brock, C. Lefurgy, A.J. Drake, L. Pesantez, T. Gloekler, J.A. Tierno, P. Bose, A. Buyuktosunoglu, Introducing the adaptive energy management features of the POWER7 chip. IEEE Micro **31**(2), 60–75 (2011)
29. J. Tschanz, N.S. Kim, S. Dighe, J. Howard, G. Ruhl, S. Vangal, S. Narendra, Y. Hoskote, H. Wilson, C. Lam, M. Shuman, C. Tokunaga, D. Somasekhar, S. Tang, D. Finan, T. Karnik, N. Borkar, N. Kurd, V. De, Adaptive frequency and biasing techniques for tolerance to dynamic temperature-voltage variations and aging, in *IEEE International Solid-State Circuits Conference. ISSCC 2007. Digest of Technical Papers* (2007), pp. 292–604
30. S. Hoppner, H. Eisenreich, S. Henker, D. Walter, G. Ellguth, R. Schuffny, A compact clock generator for heterogeneous GALS MPSoCs in 65-nm CMOS technology. IEEE Trans. Very Large Scale Integr. (VLSI) Syst. **21**(3), 566–570 (2013)
31. LEON3. http://www.gaisler.com/cms/
32. ARM Cortex-M3 technical reference manual, Rev. r1p1 (2006)
33. TSIM iss. http://www.gaisler.com/index.php/products/simulators/tsim
34. A. Rajendiran, S. Ananthanarayanan, H.D. Patel, M.V. Tripunitara, S. Garg, Reliable computing with ultra-reduced instruction set co-processors, in *49th ACM/EDAC/IEEE Design Automation Conference (DAC)* (2012), pp. 697–702
35. K.A. Bowman, J.W. Tschanz, N.S. Kim, J.C. Lee, C.B. Wilkerson, S.L. Lu, T. Karnik, V.K. De, Energy-efficient and metastability-immune resilient circuits for dynamic variation tolerance. IEEE J. Solid-State Circuits **44**(1), 49–63 (2009)

# Chapter 4
# Procedure-Level Tolerance

**Abstract** This chapter raises further the level of error tolerance to procedure calls for scheduling different procedures on cores such that there is no timing errors. We propose a resilient hardware/software (HW/SW) architecture for shared-L1 processor clusters to combat both static and dynamic variations. We first introduce the notion of procedure-level vulnerability (PLV) to expose fast dynamic voltage variation and its effects to the software stack for use in runtime compensation. To assess PLV, we quantify the effect of full operating conditions on the dynamic voltage variation of a post-layout processor in 45 nm TSMC technology. Based on our analysis, PLV shows a range of 18–63 mV inter-corner variation among the maximum voltage droop of procedures. To exploit this variation, we propose a low-cost procedure hopping technique within the processor clusters, utilizing compile time characterized metadata related to PLV. Our results show that procedure hopping avoids critical voltage droops during the execution of all procedures while incurring less than 1% latency penalty.

## 4.1 Introduction

Given the close relationship between power and temperature, and the increased importance of variability in the future, treatment of variability during pre-silicon and post-silicon design stages is crucially important. Resilient circuit techniques suffer from power-hungry error recovery, which is expensive for a many core fabric. In contrast, our approach is applicable to clusters of simple processors and exploits the opportunity given by tightly coupled architecture to dynamically shift work from one core to another with minimal overhead. In this chapter, we proposes a resilient HW/SW method for shared-L1 processor clusters to combat both static and dynamic variations:

1. We introduce the notion of procedure-level vulnerability (PLV) to capture the effects of dynamic IR-drop. Using characterized PLV, we enable a software preventive methods that build upon well-known hardware detection/correction techniques for process variability and aging.

 
A. Rahimi et al., *From Variability Tolerance to Approximate Computing in Parallel Integrated Architectures and Accelerators*,
DOI 10.1007/978-3-319-53768-9_4

2. We propose a low-cost runtime procedure hopping that facilitates migration of procedures within a processor cluster, utilizing compile-time characterization (captured as metadata) of PLV.
3. An accurate gate-level analysis flow which leverages industrial design implementation tools and libraries to characterize IR-drop of individual procedures in the presence of variability is developed. We demonstrate our approach on a tightly coupled shared-L1 multi-core cluster, representative of a large class of multi-core architectures (e.g., GP-GPUs, programmable multimedia accelerators). Full post place-and-route (P&R) results in 45 nm TSMC technology confirm that the procedure hopping technique avoids the critical IR-drop during the execution of all procedures while incurring less than 1% latency penalty.

## 4.2 Variation-Tolerant Processor Clusters Architecture

% In this section, we describe the architectural detail of proposed variation-tolerant processing cluster. These clusters are the essential parallel components of many core fabrics, e.g., NVIDIA Fermi [1] features 512 CUDA processors organized into 16 groups of processing cluster. In our implementation, each cluster consists of sixteen 32-bit in-order RISC cores compliant with the SPARC V8 architecture, an intra-cluster shared level-one instruction cache (shared-L1 I$) [2], an on-chip tightly coupled data memory (TCDM), two fast logarithmic interconnections [3] for both instruction and data sides, and a hardware synchronization handler module (SHM). The shared-L1 I$ for the MIMD cluster can achieve better performance, up to 60%, than the private I$ per core [2]. On the data side, a multi-ported, multi-banked, level-one TCDM is directly connected to the interconnect. The number of memory ports is equal to the number of banks to have concurrent access to different memory locations. The logarithmic interconnection is composed of mesh-of-trees networks to support single-cycle communication between processors and memories in L1-coupled processor clusters [3]. When a read/write request is brought to the memory interface, the data is available on the negative edge of the same clock cycle, leading to two clock cycles latency for a conflict-free TCDM access. The SHM acts as an extra slave device of the logarithmic interconnect to coordinate and synchronize cores for accessing shared data on TCDM [2].

All components of the cluster work with the same frequency (memories with a 180° phase shift) decided by DFS, while only the voltage of cores is isolated by the fast level shifters thus enabling core-level dynamic VDD-hopping [4, 5]. The VDD-hopping uses three voltages provided by external DC-DC converters (no need of on-chip inductor and charge pump) to control the local voltage of the core based on the core's delay variation. To hop between three supply voltages, a device called power supply selector (PSS) is necessary. The VDD-hopping utilizes an efficient voltage transition which allows changing the supply voltage following a controlled ramp, limiting wide current variations, avoiding any supply voltage under- or over-shoot and current flowing from one source to another [5]. Silicon results of a 65 nm

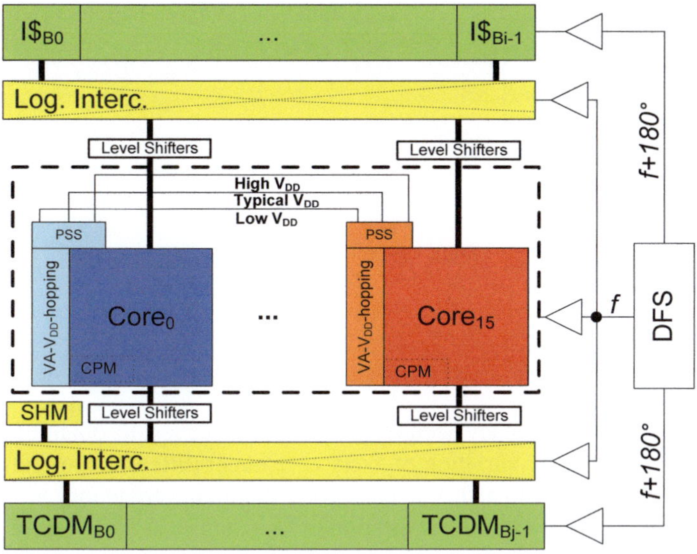


**Fig. 4.1** Variation-tolerant processor cluster

test-chip indicate that the core does not need to be stopped during VDD-hopping thanks to smooth, and fast voltage transitions (less than 100 ns), with no under-shoot or over-shoot [6]. The hopping unit and its power switches are fully integrated and are 20× smaller than the integrated buck-boost DC-DC converter [6]. As shown in Fig. 4.1, the level shifter standard cells are utilized in the back-end with a fine-grain multi-VDD design flow; each the high-to-low/low-to-high level shifter imposes only 12 ps/42 ps delay [7] (262 nW/43 nW average leakage power) for a load of fan-out-of-4, thus enabling single-cycle communications between cores and TCDM/shared-L1 I$.

## *4.2.1 Variation-Aware VDD-Hopping*

To observe the effect of process parameters variation on frequency of individual cores within a cluster, we have accurately analyzed how critical paths of each core are affected, considering the back-end details implementation of cores. Each core has been optimized during synthesis and P&R individually with a target frequency constraint of 830 MHz, then a bottom-up synthesis approach is leveraged to form the physical implementation of the cluster. After parasitic extraction, in the sign-off stage, the process parameters are varied based on die-to-die and within-die characterized process parameters variations of 45 nm TSMC models, derived from the first-level process created by principal component analysis. These standard industrial libraries and design process are supported by the state-of-the art commercial tools [7], thus

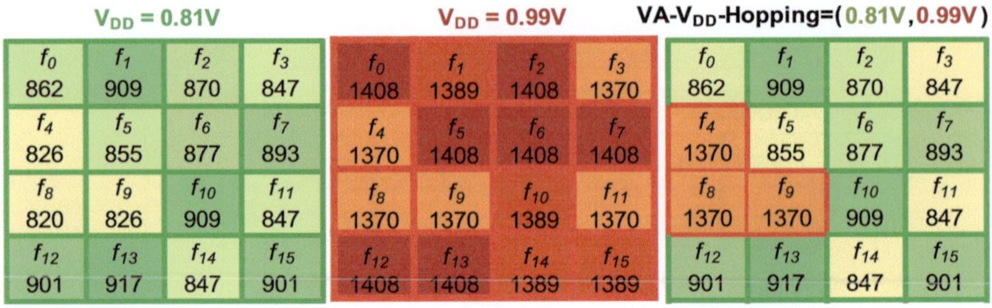


**Fig. 4.2** Frequency (MHz) variation of a 16-core cluster due to the process parameters variations under different voltages

the calculated cores' frequency accurately reflect the true results obtained in silicon. The maximum frequency variation of every core under different operating voltages is shown in Fig. 4.2. Within a cluster, each cores maximum frequency varies significantly due to increasing within-die variations. For instance, at 0.81 V, three cores (f4, f8, f9) of out of 16-core cannot meet the design time target frequency of 830 MHz.

To cope with this frequency variation problem there are three solutions: (i) limiting the frequency of cluster by the slowest core (f8 = 820 MHz); (ii) disabling the slowest cores and clocking the cluster with the next slowest core (f4 = 826 MHz); (iii) running each core at its maximum frequency independently. All these solutions impose non-negligible performance penalty; the first and second solutions directly diminish the throughput of cluster, and the last solution needs extra latency for synchronization of cores with different clock frequencies. Synchronization across multiple clock frequency islands increases the latency of interconnection which its performance impact can be as high as the cache miss.

On the other hand, we consider a core-level VDD-hopping [12] for tuning the voltage of each core individually to compensate the impact of process variation. For instance, Fig. 4.2 shows that all cores of the same cluster meet the target frequency of 830 MHz when a higher VDD (0.99 V) is applied. Therefore, every core can have its own voltage domain, while all cores can work with the target frequency utilizing the fast-level shifters. The critical paths delay of every core are measured in real time by the less intrusive and low-overhead CPMs [8], hence the variation-aware VDD-hopping (VA-VDD-hopping) can accordingly tune the cores' voltage periodically at arbitrary post-silicon stages. It mitigates both process variation and even aging slows down. Consequently, the cores which are fabricated on a fast piece of silicon will work on a lower voltage than the boosted "high VDD" voltage; this not only lowers their power but delays their aging. On the contrary, slow cores will supply at higher voltages to be able to meet the target frequency. As shown in Fig. 4.2, the VA-VDD-hopping elevates the voltage of slow cores (f4, f8, f9) to 0.99 V, while the rest of cores are supplying at 0.81 V, therefore enabling the whole cluster to clock at the target frequency of 830 MHz. Note that the VA-VDD-hopping technique mitigates the within-cluster delay variations, but imposes voltage supply changes at the core-level that can affect core's aging. Therefore, to extend service life of the slow cores the ratio of stress to recovery time can be changed using core activity duty cycling techniques [9].

## 4.3 Procedure Hopping for Dynamic IR-Drop

In the previous section, we have shown that the variability-affected cluster can combat delay variation caused by the process parameter variations and aging, leveraging the real-time observers and voltage as the control knob. CPMs observe the available slack on paths, and VA-VDD-hopping controls the voltage accordingly, this detection/correction control loop is a well-suited for those variations that: (i) have a slow time constant since compensation requires several clock cycles; (ii) contain low-frequency components to avoid the frequent cost of rollback and calibration. On the other hand, fast dynamic variations, like IR-drop, that contains high-frequency component cannot be countered by a reactive detection/correction loop. They need to be anticipated and prevented.

For this type of variations, we propose a technique consisting of two major phases: design time characterization of metadata related to PLV, and runtime preventive procedure hopping. During characterization, the probability of voltage droop/rise versus various voltage (V) and temperature (T) is characterized at the level of procedures, where the problematic sequences of instructions [10, 11] exist. Therefore, the PLV is calculated for every procedure on different combinations of (V, T) of the core, then the metadata is generated as the result. The characterized metadata is attached to each procedure at the compile time, to be able to use for runtime decisions about finding the best location to run the procedure among the available (V, T)-islands within a cluster.

During runtime, the core can evaluate the PLV of every procedure just looking at the characterized metadata, and at the same time monitoring its current (V, T) using CPMs. If the calculated PLV is greater than a predefined threshold (PLV_threshold), this means that running procedure on the original core (caller) would likely cause critical IR-drops, thus the procedure hops to another core (callee) where its (V, T) is suitable for the procedure execution. As discussed in the next subsection, procedure hopping can be done remarkably fast and proactively enough thanks to the tightly coupled shared resources within a cluster.

### *4.3.1 Supporting Intra-cluster Procedure Hopping*

Here, we describe the architectural HW/SW design to support the procedure hopping within a cluster. The goal is to facilitate fast and proactive migration of procedures from a caller core to the rest of cores, without special compiler support, minimal impact on the normal execution, and reasonable memory overhead. Figure 4.3 shows the HW/SW interactions, and steps of procedure hopping of the cluster. It is shown

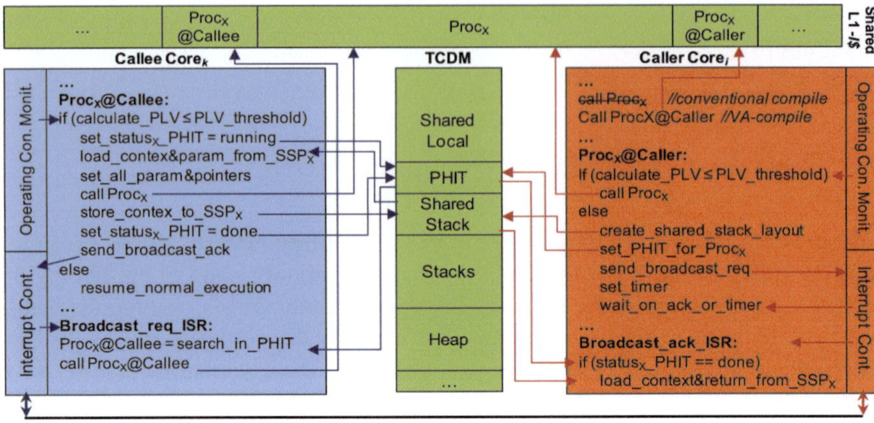


**Fig. 4.3** HW/SW collaborative architecture to support intra-cluster procedure hopping

that accessing both data and instruction is facilitated by shared TCDM and L1 I$. The shared TCDM has four regions: (i) shared local: maintains variables explicitly defined to be shared at compile time; (ii) shared stack: maintains the parameters for passing among cores; (iii) stacks: region is defined to maintain the normal stack of all 16 cores; (iv) heap: is used for dynamically allocated structures.

For every procedure, e.g., ProcX, two variation-aware procedures, ProcX@Caller and ProcX@Callee, are considered to enable runtime accesses to the characterized metadata of ProcX in the caller and callee cores respectively. The only compiler transformation is to transform "call ProcX" to "call ProcX@Caller," as shown in the code of the caller core in Fig. 4.3. Therefore, the ProcX@Caller will first run on behalf of ProcX to decide whether current (V, T) of the caller core is suitable for running ProcX or not, utilizing the metadata and reading the operating condition monitors to calculate PLV. If PLV is less than/equal to the PLV_threshold, then "call ProcX" will be executed; otherwise the procedure hopping will be applied to trigger migration of ProcX to a favor core. Once a procedure hops from the caller core to a callee core, its code is easily accessible via the shared-L1 I$ (without paying the penalty of filling a private cache), but its parameters also needed to be visible for the callee core. Therefore, a shared stack layout is created on the stack region of TCDM which is accessible via a shared stack pointer (SSP). This 36-byte shared stack layout covers the eight out registers of SPARC for passing six 32-bit parameters (%o0–%o5), a pointer to extra parameters (%o6), a return address (%o7) as well as a pointer to the return data structure. The caller core needs to copy-out the out registers and extra parameters (if available) to TCDM before migration of procedure, and then copy-in the return value or structure form TCDM to the registers after finishing execution of the migrated procedure. In our implementation, we assume that procedures do not have any global variables, and all inter-procedure communications are done through parameters passing; otherwise the caller core needs to copy-out/in all context registers (32 current registers window) to/from TCDM.

To enable the callee core to access to the data and code of a migrated procedure, a procedure hopping information table (PHIT) is considered in the shared local area of TCDM. This table simply keeps the information of a migrated procedure, including its SSP, address, and status. Every core can have up to eight nested procedure calls (the window pointer is synthesized as a 3-bit register), and only one of them can migrate, since the in-order core is a single thread core, and needs to wait for returning the result of the migrated procedure. Therefore, the 192-byte PHIT has an entry for every core which keeps the following information for a migrated ProcX: the shared stack pointer (SSPX), the address of ProcX@Callee (ADDRX), status of ProcX (STX) = {empty, waiting, running, done}.

As shown in the code of the caller core in Fig. 4.3, after filling the shared stack and PHIT, the core does a broadcast_req to inform the rest of cores about a waiting procedure for service. This broad-cast triggers an interrupt for all cores except the caller core, as potential callee candidates, which can service the waiting procedure based on their programmable priorities – the core can be programmed to ignore this interrupt or trigger it only when the core is idle. In the corresponding interrupt service routine (ISR), the callee core resumes its normal execution, and then walks through PHIT circularly, starting from its neighbor core for minimizing contention, picks up a waiting procedure to assess it. For instance, if the callee core picks up the waiting ProcX for the service, it will jump to the ADDRX, the address of ProcX@Callee. The philosophy of ProcX@Callee is like ProcX@Caller, it essentially enables the callee core to assess PLV of the ProcX based on the current operating condition of the callee core. If PLV is less than/equal to the threshold, then the callee core will access to the code and data of ProcX for executing on behalf of the caller core; otherwise the callee will resume its normal execution. Particularly, the callee core changes the STX at PHIT from waiting to running, thus the rest of cores will not pick ProcX up for the assessment – SHM device coordinates multiple concurrent accesses to PHIT. The callee core then copies-in the procedure's parameters from the shared stack via SSPX, and calls ProcX for its execution. After executing the procedure, the core copies-out the return value from register to the shared stack, sets the corresponding pointer to the return data structure (if any), sets the STX to done, and does a broad-cast_ack to inform the caller about finishing execution of ProcX.

The caller core, in the corresponding interrupt service routine of broadcast_ack, checks the STX, if it is equal to done, it then copies-in the return value and structure (if any) from the shared stack to the caller core's registers. In the time between sending a broadcast_req until receiving a broadcast_ack, the caller core can service another waiting procedure available on PHIT, or can switch to an idle mode. If the caller core does not get any ack response after a programmable timer value (e.g., 100 μs which is long enough to executing a procedure), this means that there is no better (V, T)-island (no favor core) within the cluster to prevent the voltage emergency during execution of the procedure. Therefore, the caller core sends a request to cluster's DFS controller to decrease the frequency of the whole cluster, thus lower the power density and temperature.

## 4.4 Characterization of PLV to Dynamic Operating Conditions

In this section, we demonstrate an advanced CAD flow and methodology to address variation awareness for characterization of PLV to dynamic IR-drop (we separately consider both voltage droops on VDD and voltage rises on VSS power domains), under a full range of operating conditions. It consists of two stages as shown in Fig. 4.4: (i) the design time stage which accurately analyzes the dynamic voltage droops/rises for individual procedures under full operating conditions; (ii) the compile-time stage which generates PLV metadata and corresponding variation-aware procedures. Finally, the cluster benefits from the characterized PLV at the runtime stage.

Each core of the cluster is an open-source 32-bit in-order RISC LEON3 [12] processor which is synthesized with the normal $V_{TH}$ cells of 45 nm TSMC technology, the general purpose process. The back-end optimization is performed using Synopsys IC Compiler, and then the finalized net-list and parasitics are extracted for accurate power analysis. To generate the accurate gate-level switching activity factor for the vector-based power analysis, the procedure is simulated on top of the back-end extracted net-list with timing back-annotation using Mentor Graphics ModelSim. The instantaneous power of the procedure is then analyzed under four TSMC operating conditions [7] using Synopsys PrimeTime. Providing the sign-off corner-

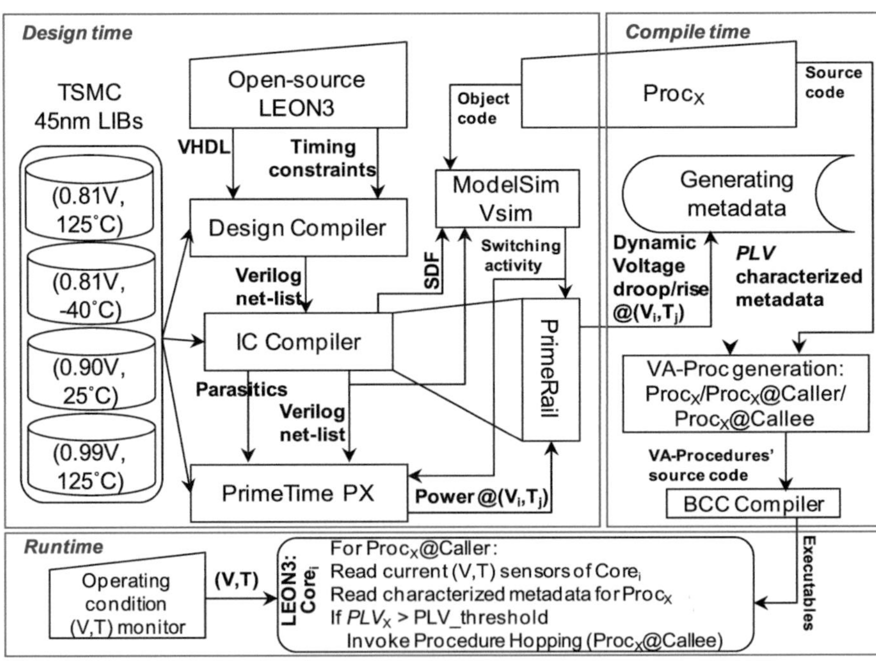


Fig. 4.4 Methodology for characterization of PLV

based instantaneous power as well as the switching activity factor enables Synopsys PrimeRail for a fine-grain, time-based rail analysis of all resistive, capacitive and inductive components of the post-P&R processor. Consequently, the inter-corner dynamic voltage droop/rise of the power rails is analyzed as the output of the design time stage.

The quantification of the $PLV_X$ (PLV of ProcX) to dynamic IR-drops defined in Eq. 4.1, where $N_X$ is the total number of clock cycles which takes to execute ProcX, and $VolEmerg_i$ indicates whether there is at least a voltage emergency at the clock $cycle_i$ or not. The voltage fluctuations of greater than 4% are viewed as voltage emergencies [10, 11] that can result in a malfunction within the processor, therefore the voltage droops/rises on VDD/VSS power rails are sampled k times during one clock cycle. The average signal activity is 70 ps, so the $k = 15$ for the target cycle time (1.2 ns), while [10, 11] sampled a second-order linear system as a model of power supply only once per cycle. The $VolEmerg_i$ is one if the maximum sampled voltage droop/rise is greater than 4% of VDD during the clock $cycle_i$. In other words, $PLV_X$ defines as the total number of cycles that have at least one voltage emergency over the total cycles for the ProcX. Intuitively, if ProcX runs without any voltage emergency, $PLV_X$ is zero; on the other hand, $PLV_X$ is one if ProcX faces at least one voltage emergency in every cycle.

$$PLVx = \frac{1}{Nx} \sum_{i=1}^{Nx} \text{VolEmerg}i$$

$$\text{VolEmerg}i = \begin{cases} 1 \text{ If Max}\{\text{drop}(t), \text{rise}(t) | t = 1, \ldots, \text{k}\} \geq \dfrac{4 \times \text{V}_{\text{DD}}}{100} \\ 0 \text{ otherwise} \end{cases} \tag{4.1}$$

$PLV_X$ is characterized for the assigned voltages of VA-VDD-hopping to various cores, {0.81 V, 0.90 V, 0.99 V} representing {fast, typical, slow} cores on a variability-affected cluster. At design time, the slow cores and fast cores are distinguished based on their maximum frequency distribution as described in Sect. 4.2, then their voltage is tuned accordingly to meet the target cluster frequency. At compile time, the characterized PLV metadata of every ProcX is attached to the two variation-aware procedures, ProcX@Caller and ProcX@Callee, to be able to runtime access to the metadata on the caller and callee cores respectively. During runtime, the discretized (V, T) operating conditions are reported by sensors thus enabling ProcX@Caller/Callee to point to the corresponding characterized PLV metadata to assess the vulnerability of ProcX at the current (V, T).

## 4.5 Experimental Results

This section shows the experimental results for embedded micro-processor AutoBench suite of benchmarks [13] characterized at the design time flow of Fig. 4.4. This section also evaluates the effectiveness of the procedure hopping technique to avoid voltage emergencies, and quantifies its latency overhead as well as the voltage droop/rise during the runtime stage. Every benchmark is a program consists of a "run" procedure for its major computation which is selected for characterization[1] and can be run on every core – the cluster is a multi-programmed environment. The inter-corner and intra-corner variations in the peak power of procedures are shown in Fig. 4.5. The corner with higher (V, T) has higher power density which imposes higher peak power. It is shown that the maximum inter-corner peak power variation is 3.5× for FIR, while the maximum of 1.28× intra-corner peak power variation occurs between IFFT and tblook procedures at (0.81 V, 125 °C). Furthermore, the maximum of 4.1× peak power variation is observed across corners and procedures, a2time at (0.81 V, −40 °C), and IFFT at (0.99 V, 125 °C). We should point out that LEON3 is a simple in-order RISC processor, thus for fast and complex cores where the stress on the power grid is much higher, we expect to see even higher power variation. Increasing the (V, T) increases the power density as well as the peak power, consequently the power network of the core highly experiences the voltage emergencies in the high-power corner. The voltage droops of running FIR on the same core but various operating corners are shown in Fig. 4.6. The core at the high-power corner (0.99 V, 125 °C) faces the maximum voltage droop of 44 and 41 mV as the average of top-100 dynamic voltage droops, which are greater than 4% of VDD (990 mV), thus these voltage droops are considered as the voltage emergencies. As opposed to the high-power corner (0.99 V, 125 °C), FIR does not face any voltage emergency at the corners with voltages of 0.90 V/0.81 V thanks to their lower power densities. The

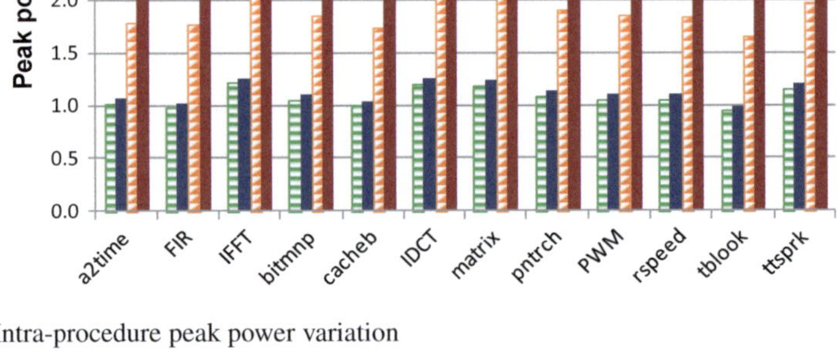


**Fig. 4.5** Intra-procedure peak power variation

[1]PLV_threshold is set at zero, since we assume that the procedures are not inherently resilient to any timing error and even a single IR-drop may cause a wrong result.

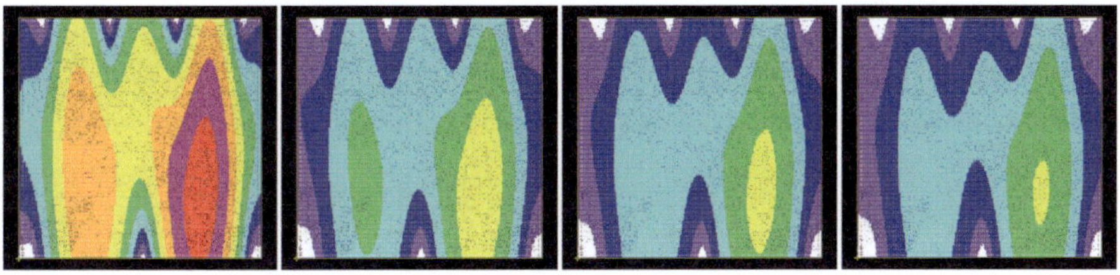

**Fig. 4.6** Voltage droop of FIR across corners: (0.99 V, 125 °C), (0.90 V, 25 °C), (0.81 V, 125 °C), (0.81 V, −40 °C), *left* to *right*

core has various power densities across the corners of Fig. 4.6 (left to right): 0.66, 0.21, 0.18, 0.16 $\mu$W/$\mu$m$^2$.

Figure 4.7 illustrates the maximum voltage droop/rise that occurs during the execution of the procedures under the four characterized operating conditions. All procedures running at cores with 0.81 V have the maximum voltage droop/rise less than 4% of VDD. Increasing the power density by switching to (0.90 V, 25 °C) causes only four procedures (IFFT, IDCT, matrix, ttsprk) to face the voltage emergencies. At the highest power corner, (0.99 V, 125 °C), most of the procedures except tblook will face either voltage droop or voltage rise greater than 4% of VDD. These results show that the procedure hopping technique can avoid the voltage emergency for all procedures by hopping them form a high-voltage (0.99 V) core to a low-voltage (0.81 V) core. Experimental results from the layout of variability-affected cluster, show that 13 low-power cores lie within a cluster of 16-core, thus providing enough callee cores to service the migrated procedures.

## 4.5.1 Cost of Procedure Hopping

Table 4.1 lists the latency overhead of involving the procedure hopping both in the caller and the callee cores. The total roundtrip overhead of the hopping a procedure from the caller core and returning the results from the callee core is 793 cycles; this is less than 1% of the total cycles needed to execute any of the characterized procedures in [13], while [14] has at least a migration overhead of transferring 1280 flits only to transfer the instructions and data from one core to another. In particular, if a procedure has a runtime of 35 K cycles, the amortized cost is only 2 and 0.2% latency penalty, in case of hopping procedure to another core, or keep running procedure on the same core respectively. This is accomplished through the advantage of shared-L1 I$ and TCDM that eliminates the penalty of filling a private storage.

Moreover, during the procedure hopping no voltage emergency can occur even at (0.99 V, 125 °C), neither in the caller nor the callee core, since the copy-in/out parameters from/to registers/TCDM does not cause any burst of activity. Consequently, the procedure hopping guarantees the voltage emergency-free migration of all procedures, fast and proactively enough.

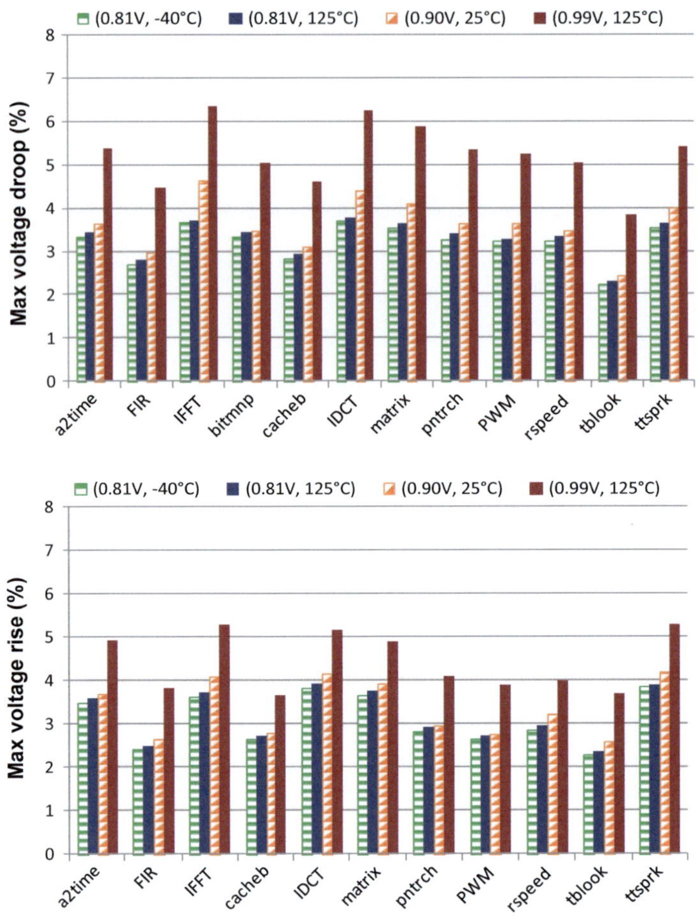


**Fig. 4.7** Percentage of the max voltage droop (*top*), and rise (*bottom*) across various corners and procedures

**Table 4.1** Latency overhead and IR-drops of procedure hopping

| | Caller hopping | Caller not hopping | Callee service | Callee no service |
|---|---|---|---|---|
| Latency | 218 cycles | 88 cycles | 575 cycles | 342 cycles |
| Max drop | 1.3% | 0.6% | 2.9% | 1.8% |

## 4.6 Chapter Summary

This chapter presents a method for predicting and preventing timing errors at the interface of procedure calls. Accordingly, we define a notion of procedure-level vulnerability (PLV) to capture fast dynamic voltage variations. Based on PLV metadata, a fully software low-cost procedure hopping technique is proposed which facilitates fast and proactive migration of procedures within a shared-L1 processor cluster. Full post-P&R results in 45 nm TSMC technology confirms that the procedure hopping avoids the voltage emergency across a variability-affected cluster, while imposing only an amortized cost of less than 1% latency for any of the characterized embedded procedures. Furthermore, the effectiveness of the variation-aware VDD-hopping technique to combat intra-cluster static variation has been demonstrated.

## References


1. Whitepaper. NVIDIA'S next generation CUDA compute architecture: Fermi (2009)
2. F. Paterna, C. Pinto, A. Marongiu, M. Ruggiero, L. Benini, Exploring instruction caching strategies for tightly-coupled shared-memory clusters, in *International Symposium on System on Chip (SoC)* (2011), pp. 34–41
3. A. Rahimi, I. Loi, M.R. Kakoee, L. Benini, A fully-synthesizable single-cycle interconnection network for shared-l1 processor clusters, in *Design, Automation Test in Europe Conference Exhibition (DATE)* (2011), pp. 1–6
4. B.H. Calhoun, A.P. Chandrakasan, Ultra-dynamic voltage scaling (UDVS) using sub-threshold operation and local voltage dithering. IEEE J. Solid-State Circuits **41**(1), 238–245 (2006)
5. S. Miermont, P. Vivet, M. Renaudin, A power supply selector for energy- and area-efficient local dynamic voltage scaling, in *Proceedings of the 17th International Workshop on Integrated Circuit and System Design. Power and Timing Modeling, Optimization and Simulation, PATMOS'07* (Springer, Berlin, 2007), pp. 556–565
6. E. Beigne, F. Clermidy, H. Lhermet, S. Miermont, Y. Thonnart, X.-T. Tran, A. Valentian, D. Varreau, P. Vivet, X. Popon, H. Lebreton, An asynchronous power aware and adaptive NoC based circuit. IEEE J. Solid-State Circuits **44**(4), 1167–1177 (2009)
7. TSMC 45 nm standard cell library release note, v 120a (2009)
8. A. Drake, R. Senger, H. Deogun, G. Carpenter, S. Ghiasi, T. Nguyen, N. James, M. Floyd, V. Pokala, A distributed critical-path timing monitor for a 65 nm high-performance microprocessor, in *IEEE International Solid-State Circuits Conference. ISSCC 2007. Digest of Technical Papers* (2007), pp. 398–399
9. F. Paterna, L. Benini, A. Acquaviva, F. Papariello, A. Acquaviva, M. Olivieri, Adaptive idleness distribution for non-uniform aging tolerance in multiprocessor systems-on-chip, in *Design, Automation Test in Europe Conference Exhibition, DATE'09* (2009), pp. 906–909
10. K. Hazelwood, D. Brooks, Eliminating voltage emergencies via microarchitectural voltage control feedback and dynamic optimization, in *Proceedings of the 2004 International Symposium on Low Power Electronics and Design. ISLPED'04* (2004) pp. 326–331
11. V.J. Reddi, D. Brooks, Resilient architectures via collaborative design: maximizing commodity processor performance in the presence of variations. IEEE Trans. Comput.-Aided Des. Integr. Circuits Syst. **30**(10), 1429–1445 (2011)

12. LEON3. http://www.gaisler.com/cms/
13. EEMBC benchmark consortium. http://www.eembc.org
14. S. Dighe, S.R. Vangal, P. Aseron, S. Kumar, T. Jacob, K.A. Bowman, J. Howard, J. Tschanz, V. Erraguntla, N. Borkar, V.K. De, S. Borkar, Within-die variation-aware dynamic-voltage-frequency-scaling with optimal core allocation and thread hopping for the 80-core teraflops processor. IEEE J. Solid-State Circuits **46**(1), 184–193 (2011)

# Chapter 5
# Kernel-Level Tolerance

**Abstract** Negative bias temperature instability (NBTI) adversely affects the reliability of a processor by introducing new delay-induced faults. However, the effect of these delay variations is not uniformly spread across functional units and instructions: some are affected more (hence less reliable) than others. For massive number of kernels executing on functional units in GPUs, we propose a preventive method to ensure the absence of NBTI-induced timing errors during GPU lifetime. This chapter presents an NBTI-aware compiler-directed very long instruction word (VLIW) assignment scheme that uniformly distributes the stress of instructions with the aim of minimizing aging of GP-GPU architecture without any performance penalty. The proposed solution is an entirely software technique based on static workload characterization and online execution with NBTI monitoring that equalizes the expected lifetime of each processing element by regenerating aging-aware *healthy kernels* that respond to the specific health state of GP-GPU. We demonstrate our approach on AMD Evergreen architecture where iso-throughput executions of the healthy kernels reduce NBTI-induced voltage threshold shift up to 49% (11%) compared to naive kernel executions, with (without) architectural support for power-gating. The kernel adaption flow takes average of 13 ms on a typical host machine thus making it suitable for practical implementation.

## 5.1 Introduction

Among various aging mechanisms, the generation of interface traps under NBTI in PMOS transistors has become a critical reliability issue in determining the lifetime of CMOS devices [1]. NBTI effects can be significant: its impact on circuit delay is about 15% on a 65 nm technology node and it gets worse in sub-65 nm nodes [2]. Nonuniform NBTI-induced performance degradation is a major concern for many-core GP-GPUs with up to 320 five-way VLIW processors. To address this issue:

1. We propose an online adaptive reallocation strategy to mitigate NBTI-induced performance degradation in GP-GPU machines. This is accomplished through a NBTI-aware compiler that uses a dynamic binary optimizer. During dynamic recompilation, the binary is optimized by customizing the kernels code with

 
A. Rahimi et al., *From Variability Tolerance to Approximate Computing in Parallel Integrated Architectures and Accelerators*,
DOI 10.1007/978-3-319-53768-9_5

respect to specific health state of GP-GPU. This technique leverages a compiler-directed scheme that uniformly distributes the stress of instructions throughout various VLIW resource slots, results in a healthy code generation that keeps the underlying GP-GPU hardware healthy.
2. We propose a fully software solution that uses static (offline) workload characterization and online availability of NBTI sensors. The dynamic binary optimizer correlates the device stress time with instructions distribution, and equalizes the expected lifetime of each processing element without any architectural modification.
3. We demonstrate our approach on AMD Evergreen GP-GPU architecture and its tool-chain to adapt kernels to the health state of GP-GPU. The throughput of our healthy kernel execution is the same as naive kernel execution (iso-throughput). In comparison with the naive kernels, our healthy kernels execution achieves a maximum 49% reduction in NBTI-induced $V_{th}$ shift over five years if GP-GPU supports power-gating during idle states. Power-gating is intrinsically protective against NBTI by providing sleep states that spare gates from stress that produces NBTI effects. In the absence of power-gating, our uniform self-healing NOP execution technique mitigates the $V_{th}$ shift by 11%. On average, the total execution time of the entire adaptation process is 13 ms on an Intel i5 CPU 2.67 GHz.

## 5.2 Device-Level NBTI Model

NBTI is an aging mechanism which manifests itself as an increase in the PMOS transistor threshold voltage ($V_{th}$) and causes delay-induced failures. NBTI is best captured by the Reaction–Diffusion (RD) model [3]. This model describes NBTI in two stress and recovery phases. NBTI occurs due to the generation of the interface traps at the Si–$SiO_2$ interface when the PMOS transistor is negatively biased ($V_{gs}$ = -$V_{dd}$) (i.e., stress phase). In the stress condition, some holes in the channel interact with the Si-H bonds in the interface which causes disassociation of Si-H bonds. The resulting hydrogen atom diffuses away and leaves positive traps in the interface. As a result, the $V_{th}$ of the transistor increases which in turn slows down the device. Equation 5.1 shows this increase in the $V_{th}$ due to stress [4]:

$$\Delta V_{th-stress} = (K_v\sqrt{t_{stress}} + \sqrt[2n]{\Delta V_{th-t0}})^{2n} \tag{5.1}$$

where $t_{stress}$ is the amount of time that PMOS transistor is under stress; $K_v$ has dependence on electrical field, temperature (T), and $V_{dd}$; $n$ is the time exponent parameter which is $1/6$ for $H_2$ diffusion; and $\Delta V_{th-t0}$ is the initial $V_{th}$ variation of PMOS at time zero.

Removing stress from the PMOS transistor ($V_{gs} = 0$) can eliminate some of the traps by diffusing back dissociative H atoms, which partially recover the $V_{th}$ shift. This is known as the recovery phase:

$$\Delta V_{th-recov} = \Delta V_{th-stress}(1 - \frac{2\xi_1 t_e + \sqrt{\xi_2 C t_{recov}}}{(1+\delta)t_{ox} + \sqrt{Ct}}) \tag{5.2}$$

where $t_{recov}$ is the time under recovery; $t_{ox}$ is the oxide thickness; $t_e$ is the effective oxide thickness; $t$ is the total time; C has temperature dependence; $\xi_1$, $\xi_2$, and $\delta$ are constants [4].

Bhardwaj et al. [5] derived a long-term cycle-to-cycle model as follows. In this model, the stress and recovery cycles can be simulated for i cycles to find the $V_{th}$ degradation. $\Delta V_{th-stress,i}$ and $\Delta V_{th-recov,i}$ are temporal changes in $V_{th}$ at the end of $i$*th* stress and recovery cycles, respectively:

$$\Delta V_{th-stress,i} = (K_v\sqrt{\alpha T_{clk}} + \sqrt[2n]{\Delta V_{th-recov,i}})^{2n} \tag{5.3}$$

$$\Delta V_{th-recov,i} = \Delta V_{th-stress,i}(1 - \frac{2\xi_1 t_e + \sqrt{\xi_2 C(1-\alpha)T_{clk}}}{(1+\delta)t_{ox} + \sqrt{CiT_{clk}}}) \tag{5.4}$$

where $\alpha$ is duty cycle or the ratio of time spent in the stress to one period of stress-recovery; $T_{clk}$ is the period of one stress-recovery cycle; and $i = t/T_{clk}$. The NBTI rate depends on many factors including process-related parameters, temperature, voltage, and workload. Here we focus on the impact of workload or $\alpha$ in the above equations. The duty cycle ($\alpha$) is controlled by the software to reduce the NBTI-induced effects.

A transistor with a larger $V_{th}$ than expected has lower drive current, and higher delay during a transition. The switching delay of a transistor can be roughly expressed as the alpha-power law:

$$\tau \propto \frac{V_{dd} L}{\mu(V_{dd} - V_{th})^{\alpha'}} \tag{5.5}$$

where $\mu$ is the mobility of carriers; $\alpha' \approx 1.3$ is the velocity saturation index; and $L$ is the channel length. Therefore, the delay variation $\Delta\tau/\tau$ can be derived as follows:

$$\Delta\tau/\tau = \frac{\Delta L}{L} + \frac{\Delta\mu}{\mu} + \frac{\alpha'}{V_{dd} - V_{th}}\Delta V_{th} \tag{5.6}$$

Considering only the effect of $\Delta V_{th}$ shift and neglecting other terms, the delay degradation $\Delta\tau$ is shown in Eq. 5.7:

$$\Delta\tau = \frac{\alpha' \Delta V_{th}}{V_{dd} - V_{th-t0}}\tau 0 \tag{5.7}$$

where $V_{th-t0}$ is the original transistor threshold voltage (at time $t_0$), and $\tau_0$ is its corresponding delay before degradation. We consider the largest $\Delta V_{th}$ to calculate the worst case delay degradation [6–9] in a circuit to assess the potential benefits of proposed NBTI mitigation techniques. In our analysis, we set all the internal node states to '0' during the stress mode to determine the worst case circuit degradation that limits the lifetime of a chip.

## 5.3 GP-GPU Architecture

We focus on the Evergreen family of AMD GP-GPUs (a.k.a. Radeon HD 5000 series), designed to target not only graphics applications but also general-purpose data-intensive applications. The Radeon HD 5870 GP-GPU compute device consists of 20 compute units (CUs), a global front-end ultra-thread dispatcher, and a crossbar to connect the global memory to the L1-caches. Every CU has access to a global memory, implemented as a hierarchy of private 8 KB L1-caches, and 4 shared 512 KB L2-caches. Each CU contains a set of 16 Stream Cores (SCs) that have access to a shared 32 KB local data storage. Within a CU, a shared instruction fetch unit provides the same machine instruction for all SCs to execute in a SIMD fashion. Finally, each SC contains five processing elements (PEs), labeled X, Y, Z, W, and T constituting an ALU engine to execute Evergreen machine instructions in a vector-like fashion. The SC has also a general-purpose registers file to support private memory. The block diagram of architecture is shown in Fig. 5.1a.

Every SC is a five-way VLIW processor capable of issuing up to five floating point scalar operations from a single very long instruction word consists primarily of five slots ($slot_X$, $slot_Y$, $slot_Z$, $slot_W$, $slot_T$). Each slot is related to its corresponding PE. Four PEs (X, Y, Z, W) can perform up to four single-precision operations separately and perform two double-precision operations together, while the remaining one (T) has a special function unit for transcendental operations. In each cycle, VLIW slots supply a bundle of data-independent instructions to be assigned to the related PEs for simultaneous execution. In an N-way VLIW processor, up to N data-independent instructions, available on N slots, can be assigned to the corresponding PEs and be executed simultaneously. Typically, this is not done in practice because the compiler may fail to find sufficient instruction-level parallelism (ILP) to generate complete VLIW instructions. On average, if M out of N slots are filled during an execution, we call the achieved packing ratio is M/N. The actual performance of a program running on a VLIW processor largely depends on the packing ratio.

### 5.3.1 GP-GPU Workload Distribution

In this subsection, we analyze the workload distribution on the Radeon HD GPUs architecture, where there are many PEs to carry out computations. As it is mentioned in Sect. 5.2, NBTI-induced degradation strongly depends on the resource utilization, which depends on the execution characteristics of the workload. Thus, it is essential to analyze how often the PEs are exercised during the runtime execution of the

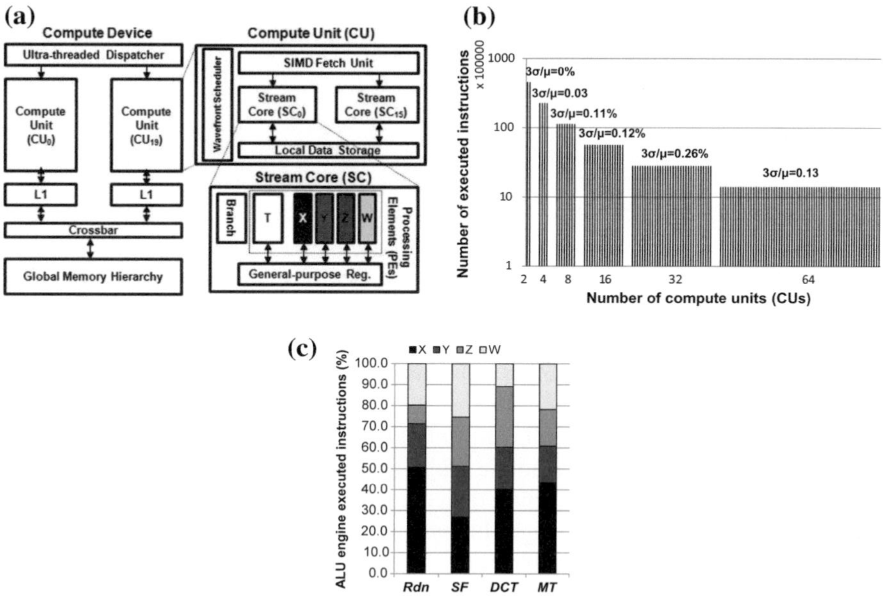


**Fig. 5.1** **a** Block diagram of the Radeon HD 5870 architecture. **b** Inter-CU workload variations for six configured compute devices. **c** Inter-PE ALU instructions distribution for various naive kernels in the HD 5870 compute device (#CUs = 20)

workload. To this end, we first monitor the utilization of various CUs (inter-CU), and then the utilization of PEs within a CU (intra-CU).

To examine the inter-CU workload variation, the total number of executed instructions by each CU is collected during a kernel execution as per a methodology described in Sect. 5.5. Figure 5.1b shows that the CUs execute almost equal number of instructions, and there is a negligible workload variation among them. We have configured six compute devices with different number of CUs, {2, 4,..., 64}, to finely examine the effect of the workload variation on a variety of GP-GPU architecture (The latest Radeon HD 5000 series, HD 5970, has 40 CUs featuring 4.3 billion transistors in 40 nm). During DCT kernel execution, the workload variation between CUs ranges from 0 to 0.26% depends on the number of physical CUs on the computation device. The DCT input kernel parameters are fixed for all configured compute devices, thus they carry out the same amount of workload—note that the total number of executed instructions per CU is inversely proportional to the number of available CUs on the compute device. Execution of all kernels listed in Sect. 5.5 confirms that the inter-CU workload variation is less than 3%, when running on the device with 20 CUs (HD 5870). This nearly uniform inter-CU workload distribution is accomplished by load balancing and uniform resource arbitration algorithms of the ultra-thread dispatcher.

Next, we examine the workload distribution among the PEs. Figure 5.1c shows the percentage of the executed instructions of ALU engine by various PEs during execution of different kernels. ALU engine here refers to four PEs ($PE_X$, $PE_Y$, $PE_Z$, $PE_W$) which are identical in their functions [10]; they differ only in the vector elements to which they write their result at the end of the VLIW. As shown, the instructions are not uniformly distributed among PEs. For instance, the $PE_X$ executes roughly half of the ALU engine instructions (50.7%) during Rdn kernel execution, while only about one quarter of the ALU engine instructions (27.1%) are executed by $PE_X$ during SF kernel execution. Execution of all kernels listed in Sect. 6 shows that seven kernels execute more than 40% of the ALU engine instructions only on $PE_X$. This nonuniform workload variation causes nonuniform aging among PEs, and exhausts some PEs more than others and shortening their lifetime. Unfortunately, this nonuniformity happens within all CUs since their workload is highly correlated together, therefore no PE throughout the entire compute device is immune from this unbalanced utilization.

Thus, root cause of nonuniform aging among PEs is the frequent and nonuniform execution of VLIW slots. For example, higher utilization of $PE_X$ implies that $slot_X$ of VLIW is occupied more frequently than the other slots. This substantiates that the compiler does not uniformly assign the independent instructions to various VLIW slots, mainly because the compiler only employs optimizations for increasing the packing ratio through finding more ILP to fully pack the VLIW slots. The VLIW processors are designed to give the compiler tight control over program execution; however, the flexibility afforded by such compilers, for instance to tune the order of instructions packing, is rarely used toward reliability improvement.

## 5.4 Aging-Aware Compilation

The key idea of an aging-aware compilation is to assign independent instructions uniformly to all slots: idling a fatigued PE and reassigning its instructions to a young PE through swapping the corresponding slots during the VLIW bundle code generation. This basically exposes the inherent idleness in VLIW slots and guides its distribution that does matter for aging. Thus, the job of dynamic binary optimizer, for K-independent instructions, is to find K-young slots, representing K-young PEs, among all available N slots, and then assign instructions to those slots. Therefore, the generated code is a "healthy" code that balances workload distribution through various slots maximizing the lifetime of all PEs. In this section, we describe how these statistics can be obtained from silicon, and how compiler can predict and thus control the nonuniform aging. The adaptation flow is illustrated in Fig. 5.2 through four steps: (1) reading aging sensors; (2) kernel disassembler, static code analysis, and calibration of predictions; (3) uniform slot assignment; (4) healthy code generation.

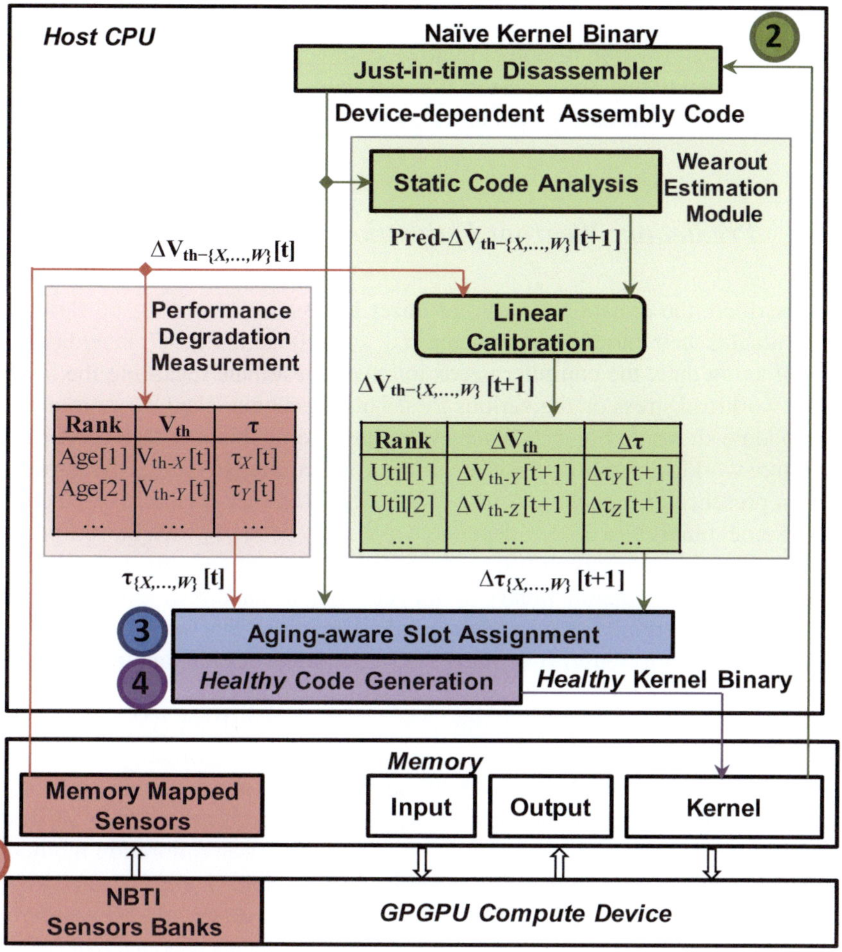


**Fig. 5.2** Aging-aware kernel adaptation flow

### *5.4.1 Observability: Aging Sensors*

The compiler needs to access the current aging data ($\Delta V_{th}$) of PEs to be able to adapt the code accordingly. The $\Delta V_{th}$ is caused by the temporal degradation due to NBTI and/or the intrinsic process variation, thus PEs even during early life of a chip might have different aging. Employing the compact per-PE NBTI sensors [11] which provide $\Delta V_{th}$ measurement with $3\sigma$ accuracy of 1.23 mV for a wide range of temperature, enables large-scale data-collection across all PEs. The performance degradation of every PE can be reliably reported by a per-PE NBTI sensor, thanks to the small overhead of these sensors. Test chips efficiently consider multiple sensors banks containing up to total 256 NBTI sensors (in 45 nm), hence the power overhead of laying out thousands of sensors would only be a few hundreds of $\mu$W at maximum, which is a small fraction of power relative to a PE [12]. The sensors support digital frequency outputs that are accessed through memory-mapped I/O by the dynamic binary optimizer in arbitrary epochs of the post-silicon measurement.

### 5.4.2 *Prediction: Wearout Estimation Module*

As described, the dynamic binary optimizer accesses to the $\Delta V_{th}$ of various PEs, and evaluates their current performance ($\tau_{\{X,...,W\}}$[t]) using Eq. 5.7. In addition to the current aging data, the compiler needs to have an estimate regarding the impact of future workload stress on the various PEs. This is accomplished by wearout estimation module shown in Fig. 5.2. Since every naive kernel binary can be considered as the future workload, code analysis techniques are required to predict the future workload in presence of branches. A just-in-time disassembler disassembles the desired naive kernel binary to a device-dependent assembly code in which the assignment of instructions to the various slots (corresponding PEs) are explicitly defined, and thus observable by the dynamic binary optimizer. Then, a static code analysis technique is applied that estimates the percentage of instructions that will be carried out on every PE in a static sense. It extracts the future stress profile, and thus the utilization of various PEs using the device-dependent assembly code. Then, the static code analysis technique predicts the future $\Delta V_{th}$ shift of PEs (Pred-$\Delta V_{th-\{X,...,W\}}$[t+1]). If the predicted $\Delta V_{th}$ of a PE is overestimated or underestimated, mainly due to the static analysis of the branch conditions of the kernel's assembly code, a linear calibration module fits the predicted $\Delta V_{th}$ shift to the observed $\Delta V_{th}$ shift, in the next adaptation period. For every PE, e.g., $PE_X$, the linear calibration module uses the simple linear regression with an explanatory variable (Pred-$\Delta V_{th-X}$[t+1]), and a dependent variable ($\Delta V_{th-X}$[t+1]). The simple linear regression fits a straight line through the set of m points (each kernel execution) in such a way that makes the sum of squared residuals of the model as small as possible. The model is developed during online measurement by observing the actual $\Delta V_{th}$ shift reported by NBTI sensors ($\Delta V_{th-X}$[t]) after each kernel execution. Therefore, the linear calibration for every PE determines the curve that best describes the relationship between expected and observed sets of $\Delta V_{th}$ data; it projects the future $\Delta V_{th}$ of PEs ($\Delta V_{th-\{X,...,W\}}$[t+1]) by minimizing the sums of the squares of deviation between observed and expected values. Finally, $\Delta V_{th-\{X,...,W\}}$[t+1] is used to calculate the future NBTI-induced performance degradation ($\Delta\tau_{\{X,...,W\}}$[t+1]).

### 5.4.3 *Controllability: Uniform Slot Assignment*

Thus far, we have described how the dynamic binary optimizer evaluates the current performance degradation (aging) of every PE ($\tau_{\{X,...,W\}}$[t]), and their future performance degradation ($\Delta\tau_{\{X,\ldots,W\}}$[t+1]) due to the naive kernel execution. Then, the compiler uses that information to perform code transformations with the goal of improving reliability, without any penalty in the throughput of code execution (maintaining the same ILP). To minimize stresses, the compiler sorts the predicted performance degradation of the slots increasingly and the aging of the slots decreasingly, and then applies a permutation to assign fewer/more instructions to higher/lower stressed slots. This algorithm for every period of adaptation [t] is shown below:

```
Degrad[1,2,3,4] = Rank_degradation_increasingly (Δτ{X,Y,Z,W}[t+1])
Age[1,2,3,4] = Rank_aging_decreasingly (τ{X,Y,Z,W}[t])
For i = 1 to 4
    Reallocate (slot (Age[i]) ← slot (Degrad[i]))
```

where slot($Degrad_{[1]}$) is the slot that will have the minimum number of instructions during the future execution of the kernel, and slot($Age_{[1]}$) is the slot that its corresponding PE has the highest aging. To take into account both initial and temporal degradations, our algorithm considers the highest aging value across the same type of PE since the lifetime of the chip is limited by the most aged component. Moreover, there is no means in the assembly code to distinguish the same type of PEs spread out among all CUs, unless the hardware architectural scheduler provides support. As a result of the slot reallocation, the minimum/maximum number of instructions is assigned to the highest/lowest stressed slot for the future kernel execution, thus uniforming the lifetime of PEs.

Execution of all examined kernels shows that the average packing ratio is 0.3, which means there is a large fraction of empty slots in which PEs can be relaxed during kernels execution. Evergreen ISA states that when a slot is empty, i.e., no instruction is specified for that slot in a VLIW bundle, the corresponding PE implicitly executes a NOP instruction [10]. Overall, our solution slips the preassigned instructions from high stressed slot, thus they will have more NOP instructions to execute instead of the stressful instructions. This reduces their total stress time and effectively decreases and thus $\Delta V_{th}$. We can assume that during a NOP execution the PE is power-gated as it invalidates the written result in the corresponding vector elements at the end of NOP execution [10]. The feasibility of single-cycle power-gating is validated by Intel through a fine-grained power-gating for a 45 nm SIMD tile [13]. Nevertheless, even in the absence of power-gating, the NOP instruction execution is self-healing that can reduce the stress time of the PE adequately. Moreover, the NOP instruction itself can be designed to highly minimize the NBTI effect [14]. We compare the benefit of a GP-GPU architecture with and without power-gating for our approach in Sect. 5.5.

Among the available software knobs to mitigate NBTI, our algorithm aims to equalize the duty cycle ($\alpha$) across all the slots. Another knob is the input pattern which is impractical to predict both in the complex workloads and circuits, thus our wearout estimation module relies on the online NBTI-induced measurement feedback through the linear calibration module for better adaptation. The proposed compiler-directed reliability approach superposes on top of all optimization performed by naive compiler and does not incur any performance penalty, since it only reallocates the VLIW slots (slips the scheduled instructions from one slot to another) within the same scheduling and order determined by the naive compiler. In other words, this dynamic binary optimizer guarantees the iso-throughput execution of the healthy kernel. It also runs fully in parallel with GP-GPU on a host CPU, thus there will be no penalty for GP-GPU kernel execution if dynamic compilation of one kernel can be overlapped with the execution of another kernel.

## 5.5 Experimental Results

Our methodology is based on AMD accelerated parallel processing (APP) software ecosystem suitable for stream applications written in OpenCL. The stream kernels are compiled into GP-GPU device-specific binaries using the OpenCL compiler tool-chain which uses a standard off-the-shelf compiler front-end (g++), as well as the low-level virtual machine framework with extensions for OpenCL as the back-end. We have implemented our dynamic binary optimizer tool using C++ leveraging AMD compute abstraction layer (CAL) APIs. CAL provides a runtime device driver library that supports code generation, kernel loading and execution, and allows applications to interact with the stream cores at the low-est-level. Multi2Sim [15] cycle-accurate simulation framework – a CPU-GPU model for heterogeneous computing targeting Evergreen ISA – is modified to collect the ALU engines statistics. We have also equipped the simulator with the NBTI sensors where our tool has access to them; in a GP-GPU chip those digitally output memory-mapped sensors can be accessed by the device management part of CAL.

The following naive binaries of AMD APP SDK 2.5 [16] kernels are run on the simulator: Reduction (Rdn), Binary Search (BSe), Haar1D (DH1D), Bitonic Sort (BSo), Fast Walsh Transform (FWT), Floyd Warshall (FW), Binomial Option (BO), Discrete Cosine Transform (DCT), Matrix Transpose/Multiplication (MT/M), Sobel Filter (SF), Uniform Random Noise Generator (URNG). Before invoking the kernel, our adaptation flow is triggered: the assembly code of the kernel using CAL APIs runtime library (aticalrt) in conjunction with NBTI sensors data is passed to the wearout estimation module, and a new code is generated that adapts the binary to the specific health state of GP-GPU. In our experiments, to keep track of aging, this flow of adaptation is also run periodically in parallel on a host CPU every hour so as to impose negligible overhead.

We consider cycle-by-cycle architectural NBTI analysis [8] in the 65 nm PTM technology with $V_{gs} = 1.2\,V$, $T = 300\,K$, and the stress statistics of the kernels execution obtained from the simulator; it is common to assume that all PMOS in a circuit degrade by the same amount [6–8]. Figure 5.3a shows the NBTI-induced $V_{th}$ degradation when executing a healthy Rdn kernel compared to the naive execution at time zero, and after one year. For this experiment, we consider a HD 5870 which is not affected by the process variability (initial inter-PE $\Delta V_{th} = 0\,mV$), and without

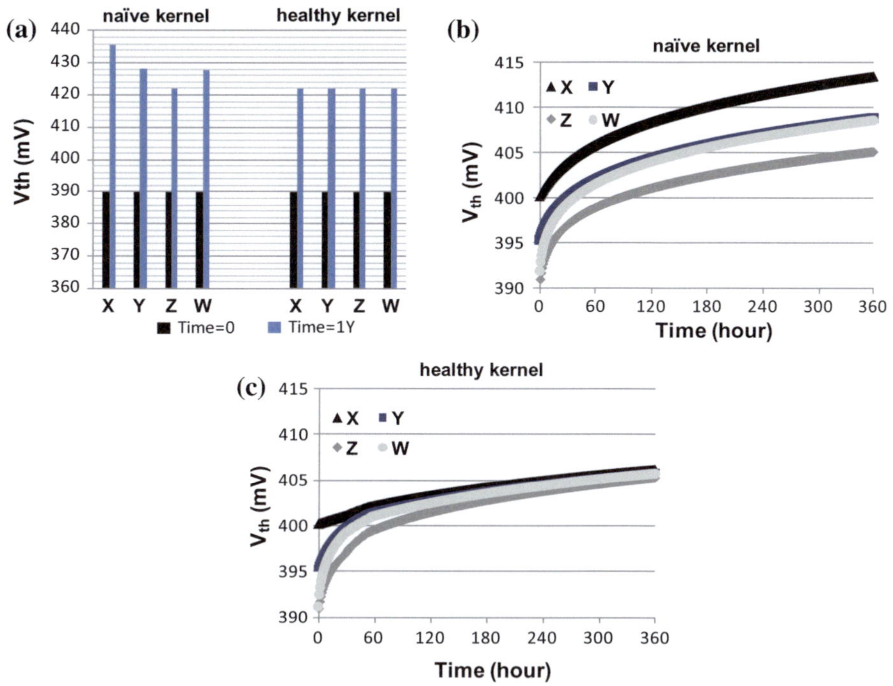


**Fig. 5.3** $V_{th}$ shift for Rdn kernel: **a** NBTI-induced for 1 year; **b** Process variation and NBTI-induced for 360 h

power-gating support. As shown in Fig. 5.3a, at time 0, all PEs have the equal $V_{th}$ since there was no stress, but after one year execution of naive Rdn, $PE_X$ has a maximum $V_{th}$ of 435 mV, because of executing 50.7% of the total ALU engine instructions (see Fig. 5.1c). However, the healthy Rdn kernel execution eliminates this nonuniformity by adapting itself every hour, and thus results in 14 mV lower $V_{th}$ shift after one year (for all PEs, $V_{th} = 421$ mV).

We also evaluate the effectiveness of the proposed approach when executing the healthy Rdn kernel on a process variability-affected HD 5870 (initial inter-PE $\Delta V_{th} = 10$ mV) and without power-gating support compared to the naive execution. Figure 5.3b shows the $V_{th}$ shift over time due to the naive kernel execution, and at the end of 360 h, there is an 8 mV $V_{th}$ variation among PEs which limits the lifetime of $PE_X$ ($V_{th-X} = 413$ mV). On the other hand, Fig. 5.3c shows that adapting the kernel periodically leads to a uniform $V_{th}$ shift among all PEs ($V_{th}$ variation is about 0.6 mV), and the maximum $V_{th}$ shift is 406 mV at the end of 360 h – with power-gating support it further reduces to 402 mV.

Indeed, the benefit of our technique is further pronounced for a larger time scale. Figure 5.4 shows the reduction in $\Delta V_{th}$ over five years execution of healthy kernels with and without power-gating support of GP-GPU architecture. In comparison with the naive execution of kernels, GP-GPU with power-gating achieves a maximum

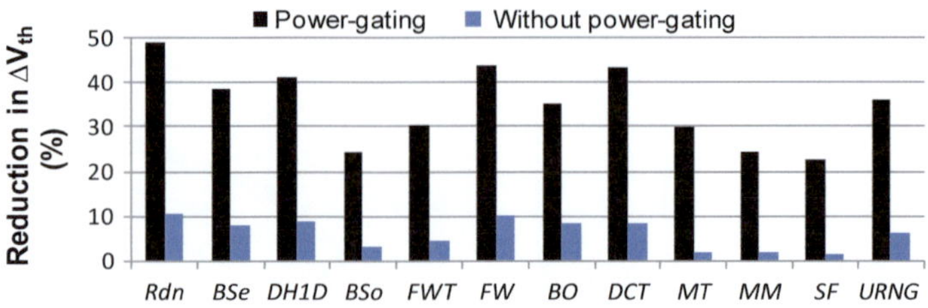


**Fig. 5.4** Reduction in $V_{th}$ due to the healthy kernels execution compared to naive kernels for 5 years

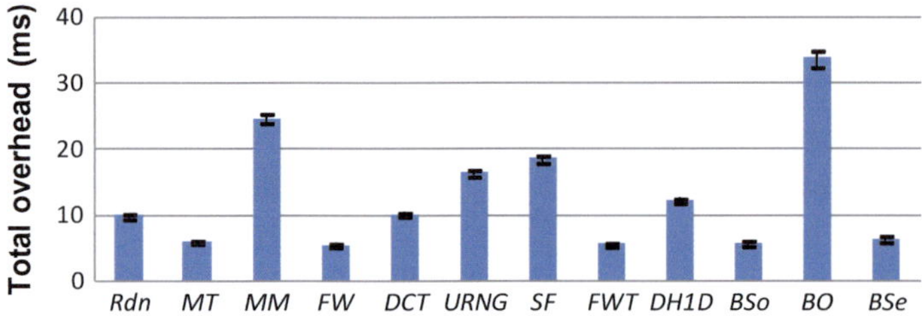


**Fig. 5.5** Total execution time of adaptation process

49% reduction in $\Delta V_{th}$, while without power-gating the self-healing NOP execution provides a maximum of 11% reduction in $\Delta V_{th}$. Since during power-gating the circuits are in the sleep state their aging mechanism are recovered quickly as derived in [17]. On average, compared to the naive kernels, the execution of healthy kernels reduces $\Delta V_{th}$ by 34 and 6% in the presence and absence of power-gating supports respectively. Furthermore, the impact of our technique is higher if we consider the local temperature reduction due to idleness and power-gating.

The total execution time of the proposed adaptation flow is measured. Figure 5.5 shows the average execution time of the entire process, starting from disassembler up to the healthy code generation. It also shows the fastest and slowest execution we measure, as error bars. More than 95% of execution time is spent through the kernel disassembly using online CAL APIs, so the assembly code can be cached for faster iterations in future adaptation. The uniform slot assignment algorithm always runs below 2 K cycles for all kernels, and the static code analysis is done between 220–900 K cycles depend to the size of kernel. Overall, the total execution time is bounded by 35 ms, and on average 13 ms on a host machine with an Intel i5 CPU 2.67 GHz.

## 5.6 Chapter Summary

This chapter presents a method for predicting and preventing the NBTI-induced timing errors at the kernel-level executing on GP-GPUs. Although the workload distribution among Compute Units (CUs) of GP-GPU is nearly uniform, its Processing Elements (PEs) suffer from nonuniform VLIW distribution. To mitigate the impacts on lifetime uncertainty and unbalancing among the PEs, an online adaptive VLIW reallocation strategy is proposed that leverages a compiler-directed scheme to uniformly distribute the stress of instructions throughout various VLIW slots. This technique periodically regenerates healthy codes that heal over GP-GPU aging. Compared to the naive kernels, the execution of healthy kernels not only imposes 0% throughput penalty but also reduces $\Delta V_{th}$: up to 49%(11%) and on average 34%(6%) in presence(absence) of architectural power-gating supports. On average, the total execution time of the adaption process is 13 ms.

## References


1. G. Chen, M.-F. Li, C.H. Ang, J.Z. Zheng, D.-L. Kwong, Dynamic NBTI of p-MOS transistors and its impact on mosfet scaling. Electron Device Lett. IEEE **23**(12), 734–736 (2002)
2. K. Bernstein, D.J. Frank, A.E. Gattiker, W. Haensch, B.L. Ji, S.R. Nassif, E.J. Nowak, D.J. Pearson, N.J. Rohrer, High-performance cmos variability in the 65-nm regime and beyond. IBM J. Res. Develop. **50**(4.5), 433–449 (2006)
3. S. Ogawa, N. Shiono, Generalized diffusion-reaction model for the low-field charge-buildup instability at the Si-Sio2 interface. Phys. Rev. **51**(7), 4218–4230 (1995)
4. W. Wang, S. Yang, S. Bhardwaj, S. Vrudhula, F. Liu, Y. Cao, The impact of NBTI effect on combinational circuit: modeling, simulation, and analysis. IEEE Trans. Very Large Scale Integr. (VLSI) Syst. **18**(2), 173–183 (2010)
5. S. Bhardwaj, W. Wang, R. Vattikonda, Y. Cao, S. Vrudhula, Predictive modeling of the NBTI effect for reliable design, in *Custom Integrated Circuits Conference, 2006. CICC '06* (IEEE, New Jersey, 2006), pp. 189–192
6. A. Tiwari, J. Torrellas, Facelift: hiding and slowing down aging in multicores, in *2008 41st IEEE/ACM International Symposium on Microarchitecture, 2008. MICRO-41* (2008), pp. 129–140
7. U.R. Karpuzcu, B. Greskamp, J. Torrellas, The bubblewrap many-core: popping cores for sequential acceleration, in *42nd Annual IEEE/ACM International Symposium on Microarchitecture, 2009. MICRO-42* (2009), pp. 447–458
8. T.-B. Chan, J. Sartori, P. Gupta, R. Kumar, On the efficacy of nbti mitigation techniques, in *Design, Automation Test in Europe Conference Exhibition (DATE)* (2011), pp. 1–6
9. F. Oboril, M.B. Tahoori, Extratime: modeling and analysis of wearout due to transistor aging at microarchitecture-level, in *2012 42nd Annual IEEE/IFIP International Conference on Dependable Systems and Networks (DSN)* (2012), pp. 1–12
10. AMD Evergreen Family Instruction Set Architecture (2011)
11. P. Singh, E. Karl, D. Sylvester, D. Blaauw, Dynamic nbti management using a 45 nm multi-degradation sensor. IEEE Trans. Circuits Syst. I Regul. Pap. **58**(9), 2026–2037 (2011)
12. P Singh, E. Karl, D. Blaauw, D Sylvester, Compact degradation sensors for monitoring NBTI and oxide degradation. IEEE Trans. Very Large Scale Integr. (VLSI) Syst. **20**(9), 1645–1655 (2012)

13. H. Kaul, M.A. Anders, S.K. Mathew, S.K. Hsu, A. Agarwal, R.K. Krishnamurthy, S. Borkar, A 300mV 494GOPS/W reconfigurable dual-supply 4-way SIMD vector processing accelerator in 45nm CMOS, in *2009 IEEE International Solid-State Circuits Conference, 2009. Digest of Technical Papers. ISSCC* (2009), pp. 260–261
14. F. Firouzi, S. Kiamehr, M.B. Tahoori, NBTI mitigation by optimized NOP assignment and insertion, in *Design, Automation Test in Europe Conference Exhibition (DATE)* (2012), pp. 218–223
15. Multi2sim: A Heterogeneous System Simulator. https://www.multi2sim.org/
16. AMD app SDK v2.5. http://www.amd.com/stream
17. A. Calimera, E. Macii, M. Poncino, NBTI-aware power gating for concurrent leakage and aging optimization, in *Proceedings of the 2009 ACM/IEEE International Symposium on Low Power Electronics and Design, ISLPED '09* (ACM, New York, NY, USA, 2009), pp. 127–132

# Chapter 6
# Hierarchically Focused Guardbanding

**Abstract** This chapter proposes a learning-based method for modeling of variation-induced timing errors in functional units. The model takes into account PVT variations and device aging (PVTA), clock frequency, and the physical details of placed-and-routed (P and R) functional units for instruction and kernel levels adaptation. Using this model and PVTA monitoring circuits, we propose *hierarchically focused guardbanding* (HFG) as a method to adaptively prevent PVTA-induced timing errors. We demonstrate the effectiveness of HFG on GPU architecture at two granularities of observation and adaptation: (i) fine-grained instruction-level; and (ii) coarse-grained kernel-level. Using coarse-grained PVTA monitors with kernel-level adaptation, the throughput increases by 70% on average. By comparison, the instruction-by-instruction monitoring and adaptation enhances throughput by a factor of 1.8×–2.1× depending on the configuration of PVTA monitors and the type of instructions executed in the kernels. This chapter presents a supervised machine learning method for reducing guardband as our last method for predicting and preventing the timing errors.

## 6.1 Introduction

Several efforts focused on online error detection and correction [1–4]. These detection and correction mechanisms do not tie to any characterized modeling, thus suffer from lack of correlation between the occurred errors and the sources of variations. This limits their usage for prediction of the timing errors and their root causes at the upper layers for better decision and appropriate adjustment. Thus, improve modeling is needed to connect the timing errors with the sources of variability for better prediction. The model should be coupled with adaptive resource management to proactively prevent the timing error by applying a *focused* guardbanding. This chapter makes the following contributions in this regard:

1. We provide a new high-level model for timing error rate (TER) of various integer as well as floating-point functional units that is derived using accurate industrial-strength tools and calibration flows validated in real silicon. This model yields the TER of microarchitectural functional units as a function of clock frequency and



A. Rahimi et al., *From Variability Tolerance to Approximate Computing in Parallel Integrated Architectures and Accelerators*,
DOI 10.1007/978-3-319-53768-9_6

the amount of PVT variations and Aging (PVTA). Section 6.2 describes the model that can be used both online and offline. Online, it provides a model-based rule to derive guardband from the PVTA sensor readings. Offline, it enables design time analysis to identify vulnerable functional units at a given amount of PVTA variations. The model is publicly available for download at [5].
2. We introduce the notion of hierarchically focused guardbanding (HFG) in Section IV to adaptively mitigate PVTA variations. HFG is guided by online utilization of the model, and enables a focused adaptive guardbanding in view of monitors, observation granularity, and reaction times.
3. We demonstrate the effectiveness of HFG using the proposed model on GPU pipeline at two distinct granularities. HFG enhances the throughput of kernels, on average by 70%, employing coarse-grained PVTA monitors and applying adaptive guardbanding at granularity of kernel-level. The finer granularity of instruction-level monitoring and adaptation achieves 1.8×–2.1× throughput improvements depending on the PVTA monitors configuration and the type of instructions executed within the kernels. Section 6.4 details the results.

## 6.2 Timing Error Model for PVTA

### *6.2.1 Analysis Flow for Timing Error Extraction*

To build a parametric model for timing errors, we rely on design time analysis that yields the TER of individual Functional Units (FUs) as a function of clock period ($t_{clk}$) and the amount of PVTA variations. We have analyzed a wide range of FUs, listed in [5], that are being used in a rich GPU pipeline, including 10 32-bit integer FUs as well as 15 single precision floating-point FUs fully compatible with the IEEE 754 floating-point standard. The floating-point FUs also cover the transcendental operations, thus act as the special FUs in the GPU pipeline to support sin, cosine, reciprocal, and square root instructions. FUs are selected from Synopsys DesignWare, a library of functions for computational circuits in high end ASICs. The speed optimized architectures have been selected for FUs in conjunction with tight synthesis and physical optimizations for timing closure. FUs have been synthesized for TSMC 45 nm target, the general purpose process. The front-end flow with normal $V_{th}$ cells uses Synopsys Design Compiler with the topographical features enabled and Synopsys IC Compiler for the backend as shown in Fig. 6.1 and Table 6.1.

For each $FU_i$ working with $t_{clk}$ and a given PVTA variations, timing error rate (TER) is defined in Eq. 6.1:

$$\text{TER}\,(\text{FU}_\text{i}, \text{t}_\text{clk}, \text{V}, \text{T}, \text{P}, \text{A}) = \frac{\sum \text{CriticalPaths}\,(\text{FU}_\text{i}, \text{t}_\text{clk}, \text{V}, \text{T}, \text{P}, \text{A})}{\sum \text{Paths}\,(\text{FU}_\text{i})} \times 100 \tag{6.1}$$

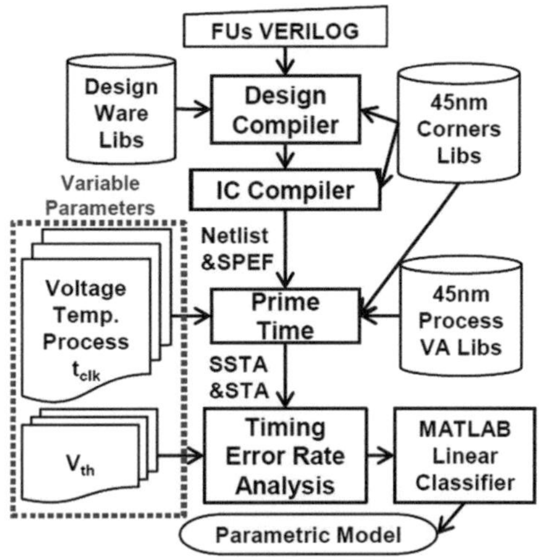


**Fig. 6.1** Timing error analysis flow for model extraction

**Table 6.1** Analysis flow: tools and parameters

| Stage | Tools/libs | Version/details |
|---|---|---|
| Front-end | Design compiler | E-2010.12-SP5 |
| Back-end | IC compiler | E-2010.12-ICC-SP5 |
| Sign-off | PrimeTime VX | F-2011.06-SP3 |
| Libraries | 45 nm GS TSMC | Variation aware (v. 110d) |
| Linear classifier | MATLAB | Discriminant analysis (v. R2011b) |

where CriticalPaths are those paths with a negative slack that cannot meet the setup-time of flip-flops with the clock period of $t_{clk}$ under certain PVTA variations, and $\Sigma$Paths is the total number of paths in FUi. After the back-end optimizations, during the sign-off, we calculate TER by analysis of FU PVTA parameter variations as follows:

**Dynamic variations**: The full industrial temperature range of 0–120 °C, and voltage range of 0.88–1.1 V are considered by utilizing six 45 nm TSMC characterized sign-off corners by changing these parameters at the resolution of 10 °C and 0.01 V, respectively. To do this, we use the voltage-temperature scaling features of Synopsys PrimeTime for the composite current source approach of modeling cell behavior. Then, at each pair of the voltage and temperature, we use static timing analysis (STA) to analyze the critical paths.

**Process variation**: The device parameters are varied from die-to-die (D2D) as well as within-die (WID), and then Statistical STA (SSTA) is used to report delay variation of each path. To perform an accurate design time SSTA, we employ the variation-aware timing analysis engine of Synopsys PrimeTime VX, using process parameters of 45 nm variation-aware TSMC libraries [6] derived from first-level process parameters by principal component analysis (PCA). PCA is a mathematical procedure that simplifies a data set by transforming a number of correlated parameters into a smaller number of uncorrelated parameters. Based on [7], the process parameters are varied as normal distributions with zero mean and standard deviations of $\sigma_{D2D} = 5\%$ and $\sigma_{WID} \in [0, 9.6\%]$. Therefore, we change the process variation components and examine its induced delay variation with a given set of accurate variability models from commercial libraries. These are more accurate and realistic than commonly used "in-house models" extracted from predictive technology models.

**Aging**: Two major mechanisms that induce progressive slowdown are NBTI and HCI, these effects manifest as voltage threshold ($V_{th}$) shift and gradually slower the critical paths. The delay of critical paths under various dynamic and process parameter variations is reported by STA and SSTA. To analyze the effect of aging on those paths, their $V_{th}$ is increased, and then their aging-induced delay variation is calculated using the alpha-power law. The $V_{th}$ is increased with steps of 25 mV and up to 100 mV which can occur over years of stress [8].

Considering the full permutation of PVTA parameters variations, the effects of variability on the delay of a FU is finely captured for its entire lifetime. To observe how this variability can be compensated by adaptive clocking, the $t_{clk}$ is changed from 0.2 to 5.0 ns. Then, TER Analysis module (Fig. 6.1) calculates TER based on $t_{clk}$ and the amount of PVTA variations using Eq. 6.1. Consequently, the calculated TER function of the five variables (summarized in Table 6.2) is input to a parametric linear classifier for model generation.

### *6.2.2 Parametric Model Fitting*

We present a parametric model at the level of FU that relates PVTA parameters variation and tclk to TER, thus enables higher level simulation and adaptiveness. To quantify the impact of timing error on the quality of service at the application-level,

**Table 6.2** PVTA and clock parameters

| | Start point | End point | Step | # of points |
|---|---|---|---|---|
| Voltage | 0.88 V | 1.10 V | 0.01 V | 23 |
| Temperature | 0 °C | 120 °C | 10 °C | 13 |
| Process ($\sigma_{WID}$) | 0% | 9.6% | 3.2% | 4 |
| Aging ($\Delta V_{th}$) | 0 mV | 100 mV | 25 mV | 5 |
| $t_{clk}$ | 0.2 ns | 5.0 ns | 0.2 ns | 25 |

**Table 6.3** Classes of TER

| TER = 0% | 33% > = TER >0% | 66%> = TER >33% | 100%> = TER >66% |
|---|---|---|---|
| $\text{Class}_0$ ($C_0$) | $\text{Class}_{\text{Low}}$ ($C_L$) | $\text{Class}_{\text{Medium}}$ ($C_M$) | $\text{Class}_{\text{High}}$ ($C_H$) |

we define four classes based on the magnitude of TER shown in Table 6.3. A higher TER implies higher number of violated critical paths, thus lower application-level quality of service. If a TER is classified as $C_0$, it means that all paths of FU meet the timing requirement; on the contrary, more than 66% of the paths (and up to 100%) are failed if a TER is classified as $C_H$. Hence, this classification covers various application-specific requirements on computational accuracy: $C_0$ for error-intolerant applications (e.g., general purpose applications), and $C_L$, $C_M$, $C_H$ for error-tolerant applications (e.g., probabilistic applications [9]) where the acceptance threshold of TER is specified according to the target quality of service of applications. We define X as a matrix of numeric predictor values [$t_{clk}$ V T P A]. Each column of X represents one variable, and each row represents one observation. Y is defined as a numeric vector, and each row of Y represents the classification of the corresponding row of X. A linear parametric classifier, called discriminant analysis, is used to create a discriminant classification based on the input variables (predictors) X and output (response) Y. Thus, the model enables mapping of the five input variables to one of the four defined classes. The discriminant analysis assumes X has a Gaussian mixture distribution. To train the classifier, the fitting function estimates the parameters of a multivariate Gaussian normal distribution for each class. After training, the classifier produces the following:

- $M\mu$ is a matrix of class means of size K-by-P, where K is the number of classes, and P is the number of predictors. Each row of $M\mu$ represents the mean of the multivariate normal distribution of the corresponding class.
- $M\sigma$ is a P-by-P matrix, the between-class covariance, where P is the number of predictors.
- Mp represents the prior probabilities for each class. Mp is a numeric positive vector of size 1-by-K representing the frequency with which each element occurs.

For each FU, the matrix of numeric predictor values, X, has a size of 149,500 ($25\times23\times13\times4\times5$)-by-5, as each row represents one permutation of the parameters summarized in Table 6.2. Every row of Y depicts the characterized class of the corresponding row of X, determined by the TER Analysis module. The space of X values divides into regions where a classification Y is a particular value. The regions are separated by straight lines for the linear discriminant analysis. Feeding X and Y to the classifier $M\mu$, $M\sigma$, and Mp are generated. The matrices for the floating-point adder ($FP_{add}$) are shown below:

$$\mathrm{M}_{\mu}=\begin{pmatrix}1.15\text{E+00} & 9.97\text{E-01} & 5.85\text{E+01} & 4.67\text{E+00} & 3.48\text{E+01}\\ 8.38\text{E-01} & 9.84\text{E-01} & 6.49\text{E+01} & 5.04\text{E+00} & 4.09\text{E+01}\\ 8.36\text{E-01} & 9.71\text{E-01} & 6.15\text{E+01} & 4.85\text{E+00} & 3.89\text{E+01}\\ 4.65\text{E-01} & 9.83\text{E-01} & 6.13\text{E+01} & 4.92\text{E+00} & 4.00\text{E+01}\end{pmatrix}$$

$$\mathrm{M}_{\sigma}=\begin{pmatrix}4.31\text{E–02} & \text{-2.37E-03} & 4.83\text{E-01} & 4.37\text{E-02} & 8.81\text{E-01}\\ \text{–2.37E-03} & 4.35\text{E-03} & 1.03\text{E-02} & 9.07\text{E-04} & 1.83\text{E-02}\\ 4.83\text{E-01} & 1.03\text{E-02} & 1.60\text{E+03} & \text{-1.91E-01} & \text{–3.80E-00}\\ 4.37\text{E-02} & 9.07\text{E–04} & \text{-1.91E-01} & 1.28\text{E+01} & \text{–3.37E-01}\\ 8.81\text{E-01} & 1.83\text{E-02} & \text{–3.80E+00} & \text{–3.37E-01} & 7.75\text{E+02}\end{pmatrix}$$

$$\mathrm{M}_{\mathrm{p}}=\begin{bmatrix}4.80\text{E-01} & 8.10\text{E-03} & 5.27\text{E-03} & 5.07\text{E-01}\end{bmatrix} \quad (6.2)$$

Providing these parametric matrices, a prediction method discussed in the next section can accurately classify a given set of variations and a $t_{clk}$ value to the corresponding class of timing error rate. The parametric models for the rest of FUs are detailed in [5] due to the lack space; the prefix 'FP' stands for floating-point FUs and "INT" stands for integer FUs.

### 6.2.3 TER Classification

A classification algorithm seeks to minimize the expected classification cost:

$$\hat{y}=\underset{y=1,\ldots,\mathrm{K}}{\arg\min}\sum_{K=1}^{\mathrm{K}}P'(k|x)\,C(y|k) \quad (6.3)$$

ŷ is the predicted classification; K is the number of classes; $\acute{P}(k|x)$ is the posterior probability of class k for observation x; C(y|k) is the cost of classifying an observation as y when its true class is k. By default, C(y|k) = 1 if y≈k, and C(y|k) = 0 if y = k: the cost is 0 for correct classification, else it is 1.

The posterior probability that a point x belongs to class k is the product of the prior probability and the multivariate normal density. The density function of the multivariate normal with mean $\mu_k$ (k-th row of M$\mu$) and covariance M$\sigma$ at a point x is

$$P(x|k)=\frac{1}{(2\pi\ |\mathrm{M}_{\sigma}|)^{0.5}}\exp\left(-\frac{1}{2}(x-\mu_k)^T\mathrm{M}_{\sigma}^{-1}(x-\mu_k)\right) \quad (6.4)$$

where |M$\sigma$ | is the determinant of M$\sigma$, and M$\sigma$-1 is the inverse matrix. Let P(k) represent the prior probability of class k (k-th element of Mp vector). Then the posterior probability that an observation x is of class k is

$$P'(k|x)=\frac{P(k|x)P(k)}{P(x)} \quad (6.5)$$

where P(x) is a normalization constant, the sum over k of P(x|k)P(k). Therefore, we can quantify the expected misclassification cost per observation. Suppose we have an observation, x = [$t_{clk}$ V T P A], to classify with the trained discriminant analysis classifier. The expected (average) cost of classifying the observation into class k of K classes is

$$\text{cost}(k) = \sum_{i=1}^{K} P^{'}(i|x)\, C(k|i) \tag{6.6}$$

$\acute{P}$(i|x)is the posterior probability defined in Eq. 6.5; and C(k|i) is the cost of classification as described in Eq. 6.3. Therefore, x belongs to the class k that has the lowest cost (k).

### *6.2.4 Robustness of Classification*

To ensure the robustness of our method, we calculate resubstitution error as the difference between the response training data and the predictions the classifier makes of the response based on the input training data. If the resubstitution error is high, we cannot expect the predictions of the classifier to be good. The resubstitution error is 0.02 (the fraction of the training data X that classifier misclassifies) for the $FP_{add}$. On average, for all FUs the resubstitution error is 0.036 which is very low, meaning the models classify nearly all data correctly.

The sampling data for prediction is almost always a subset of the training data set, since the resolution of the training data, depicted in Table 6.2, is much finer than the resolution of sampling sensors. In case of any out-of-sample data, for instance, a temperature sensor with resolution of 1 °C, the data can be conservatively matched to a surrounding point. However, we have obtained a full range of extra characterization points for temperature which are not used for training the model, and use these points to check if the model makes reasonable estimates for out-of-sample data. For extra characterization points with temperature range of 1–120 °C (steps of 1 °C) and with two distinct operating voltages (1.0, 1.1 V), the model makes correct estimates for 97% of out-of-sample data. The remaining 3% is misclassified to the high-error rate class (thus will have safe guardband). Note that we cannot go beyond the min/max range of the characterized points in the provided libraries [6].

## 6.3 Runtime Hierarchically Focused Guardbanding

We now describe how this model for TER can guide a control system for runtime variation-aware resource management. At design time, to ensure numerical correctness for the computed result, we need to take the worst-case variations that could display for any combination of values of PVTA parameters. Thus, TER can be

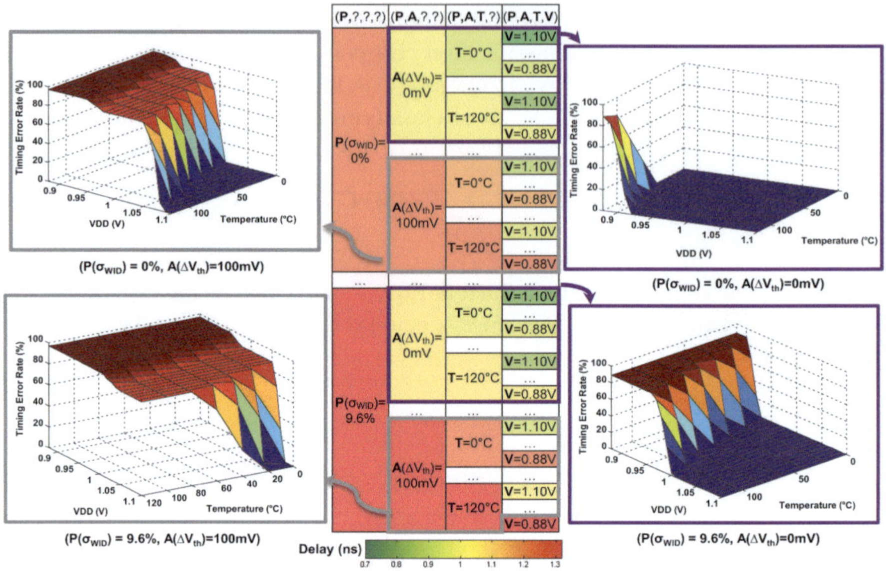


**Fig. 6.2** Delay variation and TER across extreme corners of PVTA

conservatively computed with significant uncertainty over the big cloud of possible post-silicon results. With the support of variability measurements at post-silicon fabrication, the PVTA parameters can be continuously monitored during the lifetime of the device, and consequently eliminate the conservativeness. For instance, the table in Fig. 6.2 shows that during design time the delay of the $FP_{add}$ has a large uncertainty of [0.73, 1.32 ns], since the actual values of PVTA parameters are unknown. But, immediately after fabrication this delay uncertainty is reduced to [0.73, 1.25 ns] if a process sensor reports that the adder is fabricated in a part of die with negligible WID variations. Even more, if the adder is monitored by an aging sensor, the delay uncertainty is further reduced to [0.73, 1.07 ns] when the device is fresh ($\Delta V_{th} = 0\,\text{mV}$). Having set the $t_{clk} = 0.8\,\text{ns}$, each curve in Fig. 6.2 shows how TER can change when voltage and temperature are varying at minimum/maximum process and aging conditions.

Thus, hierarchically focused guardbanding (HFG) adaptively eliminates the conservative guardband due to PVTA variations during lifetime of device. It finely focuses on a FU and reduces its timing guardband depending on the availability of distinct observers, in a hierarchical manner, started immediately after post-silicon fabrication (to compensate P), to during runtime execution (to compensate VT), and finally the entire of lifetime (to compensate A). This model-based use of PVTA readings provides a systematic way to reduce guardbands.

### 6.3.1 *Observability*

The sensor instrumentation is required as delay variation changes across extreme corners of PVTA parameters. The question is that what mix of monitors would be useful? External nonintrusive monitors reside on the same die can measure distinct parameters like voltage droop [10], and temperature fluctuation [11]. In a similar vein, CPM [12] and TRC [13] monitors whole PVT variations. On the other hand, internal in situ monitors like EDS [1], Razor [14], and NBTI sensors [8] can measure the actual delay variation of device due to PVT and aging. Figure 6.3 shows the minimum affordable $t_{clk}$ (i.e., 1/Frequency$_{Max}$) in presence/absence of various sensors for three FUs with a TER target of 0%. The sensors are sorted based on the time constant of the measured parameter, PATV: from DC component to high-frequency components. For instance, $t_{clk}$ of FP$_{add}$ can be reduced from 1.32 to 1.26 ns(a 0.06 ns guardband reduction) depends to the actual value of WID process variation reported by a process monitor (P$_{sensor}$). It can be further reduced to 1.08 ns if FP$_{add}$ is equipped with the aging as well as the process sensor (PA$_{sensors}$). Adding thermal sensor enables even 0.06 nsmore reduction to 1.02 ns(PAT$_{sensors}$). Finally, considering the full set of sensors enables decreasing $t_{clk}$ from 1.32 to 0.74 ns (a great guardband reduction of 0.58 ns) based on the measured values of variations reported by PATV$_{sensors}$. The more sensors we provide for a FU, the better conservative guardband reduction for that FU: the guardband can be reduced up to 8, 24, 28, 44%, if we equip FP$_{add}$ only with P$_{sensor}$, PA$_{sensors}$, PAT$_{sensors}$, and PATV$_{sensors}$, respectively.

As shown, this benefit is consistent across different FUs—with a shift in the worst-case guardband—even with better reduction for FP FUs (e.g., up to 47% for FP$_{exp}$ with PATV$_{sensor}$ case) due to the higher complexity of the circuit topology. Internal PVT sensors impose 1–3% area overhead [1], whereas five replica PVT sensors

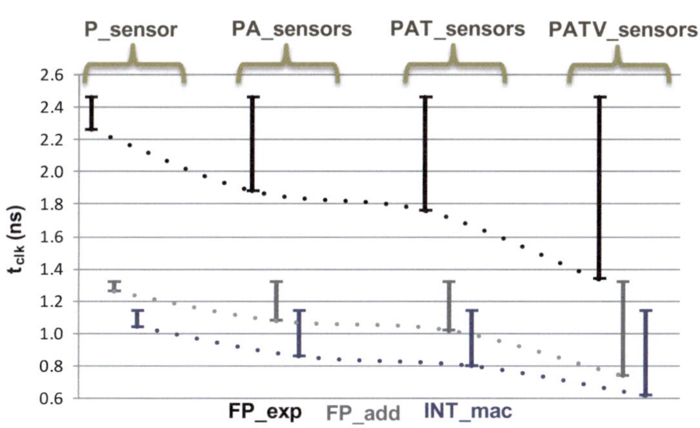


**Fig. 6.3** Hierarchical sensors for reducing guardband on $t_{clk}$

increase area of each POWER7 core by 0.2% [12, 15]. The banks of 96 NBTI aging sensors occupy less than 0.01% of the core's area [8].

### 6.3.2 Controllability

Employing any combination of PATV sensors provides online measurement of the actual parameters variations, and thus a control system can adaptively apply an appropriate guardbanding utilizing the characterized models for FUs. Among available control knobs, adaptive clock scaling using phase-locked loop (PLL) is widely utilized in resilient implementations [11, 15, 16]. Therefore, the control system tunes the clock frequency through an online model-based rule. To support fast controller's computation, the parametric model (as the outcome of the analysis flow in Fig. 6.1) generates distinct lookup tables (LUTs) for every FUs. LUTs are generated during design time for specific configuration of sensors, their resolution, and the desire target TER for FUs (target_TER). Figure 6.4 shows a full configuration of $\text{PATV}_{sensors}$ with resolutions of (3.2%, 25 mV, 20 °C, 0.04V) that support the range of variations summarized in Table 6.1. Therefore, in total 980 ($4\times5\times7\times7$) rows are required within a LUT. The parametric model fills every row of a LUT for FUi with the minimum $t_{clk}$ such that TER ($FU_i$, $t_{clk}$, $V_{row}$, $T_{row}$, $P_{row}$, $A_{row}$) < target_TER. Every LUT is stored in a dedicated 1KB SRAM to enable fast return of the 5-bit $t_{clk}$ for the corresponding values of $\text{PATV}_{sensors}$. The clock control changes the frequency based on the returned $t_{clk}$, thus reduces the guardbanding. Note that, since TER characterization in Eq. 6.1 considers the static critical paths (which might not be activated during execution of certain dynamic inputs), the model always returns an upper bound of the actual TER, thus returned $t_{clk}$ of LUTs guarantees the target_TER.

The next question to address is what type of monitoring observation granularity and what type of reacting time we need, e.g., cycle-by-cycle or tens of cycles or hundred of cycles? To analyze the effect of this choice of granularity, we apply HFG to GPU architecture at two granularities:

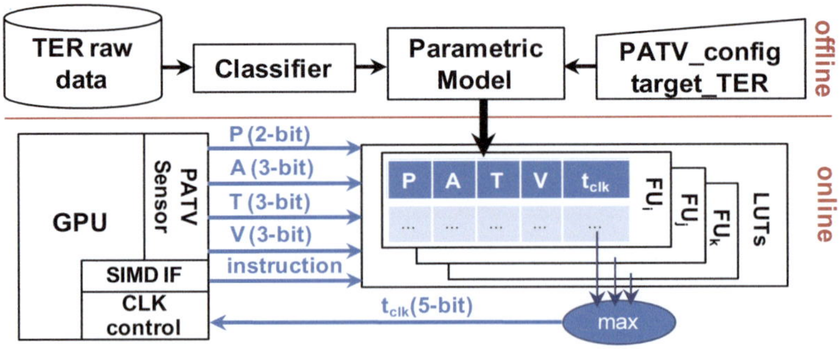


**Fig. 6.4** Online utilization of models through HFG

1. Fine-grained granularity of instruction-by-instruction monitoring and adaptation that signals of PATV sensors come from individual FUs that reside in the execution stage of GPU. The LUTs return the minimum $t_{clk}$ depending on the actual value of PATV sensors and the chain of FUs that will be activated by the fetched instruction. To support single-cycle adaptation, a fast adaptive clocking circuit [11] consisting of three PLLs is use. Each PLL is running at independent frequencies, and a multiplexer quickly switches between them in a single cycle. Therefore, the clock controller selects the highest $t_{clk}$ (safe across all activated FUs) and reduces guardband that is compatible with PATV parameters and the demands of instructions, as shown in the following algorithm:

---

$\forall$ fetched instruction$_k$
    N = #of activated FUs by instruction$_k$
    for $i$=1 to N
        $t_{clk-i}$=LUTs(FU$_i$, V, T, P, A)
    set_clock max$\{t_{clk-1}, t_{clk-1}, \ldots, t_{clk-N}\}$

---

2. Coarse-grained granularity of kernel-level monitoring uses a representative PATV sensors for the entire execution stage of GPU pipeline. The clock adaptation is applied periodically before kernel execution. The controller selects $t_{clk}$ based on current value of PATV sensors of the execution units and the chain of FUs that potentially will be activated during kernel execution (in a static sense). Since the adaptation of clock during kernel execution is prohibited, the controller considers a 5% extra margin on the reported voltage and temperature values to recover intra-kernel dynamic variations.

## 6.4 A Case Study of HFG on GPUs

We examine the effectiveness HFG on GPU architecture with the fine-grained instruction-by-instruction as well as the coarse-grained kernel-level monitoring and adaptation. We demonstrate our approach in an Evergreen-like GPU pipeline where our FUs reside in the execution stages of a processing element (PE) and benefit from the adaptive clock scaling decided by the controller of HFG. The rest of pipeline stages are assumed to support resilient circuit techniques, as both resilient processor [2] and relaxed-reliability cores [17] consider sufficient guardband in the register stage, the memory management unit, L1 instruction cache, and the interconnect. We note that the instruction fetch and decode stages are not strongly vulnerable to variations [18], thus low-cost to protect.

For GPU kernel benchmarks, we use AMD APP SDK 2.5 [19] kernels suitable for stream applications written in OpenCL. Their device-specific assembly code is extracted by AMD APP KernelAnalyzer tool for applying the instruction-by-

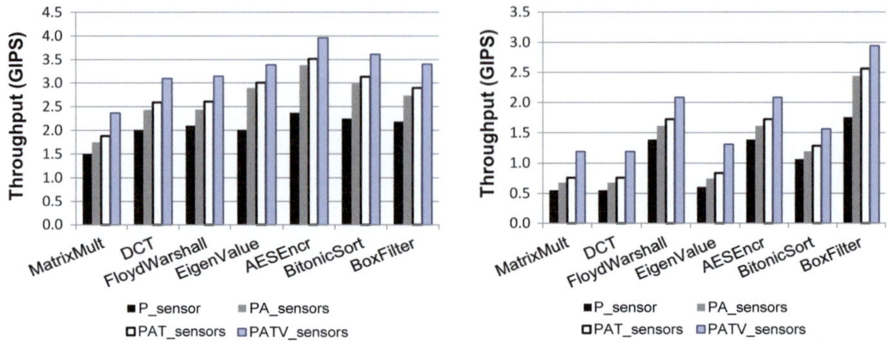


**Fig. 6.5** Maximum throughput benefit of HFG: **i** at instruction-level monitoring, the *left* figure; **ii** at kernel-level monitoring, the *right* figure

instruction and kernel-level HFG. Figure 6.5 (right) shows the maximum throughput (GIPS for a PE) of each kernel, when applying the coarse-grained kernel-level monitoring and adaptation with support of the four scenarios of PATV sensors. The results highlight two points: (a) more sensors in a PE result in a greater reduction in the guardband, and thus higher throughput for all kernels. On average, the throughput increases from 1.04 GIPS to 1.77 GIPS (70%), when the PE moves from only $P_{sensor}$ to $PATV_{sensors}$ scenario; (b) the throughput of kernel-level adaptation is limited by the slowest FU activated during the execution of the kernel. For instance, the throughput of MatrixMult, DCT, and EigenValue kernels is limited to 1.2 GIPS (with $PATV_{sensors}$), since those kernels activate $FP_{mac}$ as the slowest FU.

Figure 6.5 (left) shows the maximum throughput improvement in the instruction-by-instruction method. This method not only benefits from more sensors (60% in average), but also exploits the within-kernel opportunities for further reduction of inter-FU guardband. For example in $PA_{sensor}$ case, the throughput of AESEncr kernel is increased up to 3.4 GIPS (93% higher than MatrixMult), thanks to all its integer instructions that only activate fast INT FUs. In comparison with the kernel-level method, the instruction-by-instruction monitoring and adaptation improves the throughput by a factor of $1.8\times$–$2.1\times$ depends to the PATV sensors configuration and kernel's instructions. Of course, this fine-grained instrumentation and adaptation has a higher cost in the area.

## 6.5 Chapter Summary

This chapter presents a learning-based model and its usage for runtime variation-aware resource management as well as design time analysis of vulnerable functional units. The model takes into account process parameters, temperature and voltage operating conditions, aging, and the physical details of P and R functional units using an accurate 45 nm TSMC design and analysis flow. The model is used in a guardbanding scheme as an adaptive resource management technique to proactively prevent timing error by applying a focused guardbanding. HFG enhances the throughput of GPU kernels by 70% employing coarse-grained PVTA monitors and by applying adaptive guardbands at kernel-level. The finer granularity of instruction-by-instruction monitoring and adaptation achieves 1.8×–2.1× throughput improvements depends to the PVTA monitors configuration and the type of instructions executed within the kernels.

This methodology of deriving guardband from model-based rules, generated by supervised learning task, is also used for other applications. Interested readers may refer to [20–22]. We show that a learned model based on logistic regression can effectively predict timing errors for a given amount of guardband reduction and subject to a required bit-level *reliability specification* [20]. The same modeling approach is also used to predict the dynamic delay of various functional units based on the input workload; this shows that logistic regression can capture the dynamic path sensitization behavior under different input operands [21]. In a similar vein, a random forest method constructs a binary classifier for detecting timing errors by using input operands, computation history, and circuit toggling as the input features [22].

## References


1. K.A. Bowman, J.W. Tschanz, Nam Sung Kim, J.C. Lee, C.B. Wilkerson, S.L. Lu, T. Karnik, V.K. De, Energy-efficient and metastability-immune resilient circuits for dynamic variation tolerance. IEEE J. Solid-State Circuits **44**(1), 49–63 (2009)
2. K.A. Bowman, J.W. Tschanz, S.L. Lu, P.A. Aseron, M.M. Khellah, A. Raychowdhury, B.M. Geuskens, C. Tokunaga, C.B. Wilkerson, T. Karnik, V.K. De, A 45 nm resilient microprocessor core for dynamic variation tolerance. IEEE J. Solid-State Circuits **46**(1), 194–208 (2011). Jan
3. S. Das, D. Roberts, Seokwoo Lee, S. Pant, D. Blaauw, T. Austin, K. Flautner, T. Mudge, A self-tuning DVS processor using delay-error detection and correction. IEEE J. Solid-State Circuits **41**(4), 792–804 (2006)
4. M. Floyd, M. Ware, K. Rajamani, T. Gloekler, B. Brock, P. Bose, A. Buyuktosunoglu, J.C. Rubio, B. Schubert, B. Spruth, J.A. Tierno, L. Pesantez, Adaptive energy-management features of the IBM power7 chip. IBM J. Res. Develop. **55**(3), 8:1–8:18 (2011)
5. PVTA Models for Hierarchically Focused Guardbanding. http://mesl.ucsd.edu/site/PVTA_MODELS/models.htm
6. TSMC 45 nm Standard Cell Library Release Note, v 120a (2009)
7. S. Herbert, D. Marculescu, Characterizing chip-multiprocessor variability-tolerance, in *45th ACM/IEEE Design Automation Conference, 2008. DAC 2008* (2008), pp. 313–318
8. P Singh, E. Karl, D. Blaauw, D Sylvester, Compact degradation sensors for monitoring NBTI and oxide degradation. IEEE Trans. Very Large Scale Integr. (VLSI) Syst. **20**(9), 1645–1655 (2012)
9. P. Dubey, Recognition, mining and synthesis moves computers to the era of tera, in *Technology Intel Magazine* (2005), pp. 1–10
10. S. Pant, D. Blaauw, Circuit techniques for suppression and measurement of on-chip inductive supply noise, in *34th European Solid-State Circuits Conference, 2008. ESSCIRC 2008* (2008), pp. 134–137
11. J. Tschanz, N.S. Kim, S. Dighe, J. Howard, G. Ruhl, S. Vangal, S. Narendra, Y. Hoskote, H. Wilson, C. Lam, M. Shuman, C. Tokunaga, D. Somasekhar, S. Tang, D. Finan, T. Karnik,

N. Borkar, N. Kurd, V. De, Adaptive frequency and biasing techniques for tolerance to dynamic temperature-voltage variations and aging, in *IEEE International Solid-State Circuits Conference, 2007. ISSCC 2007. Digest of Technical Papers* (2007), pp. 292–604
12. A. Drake, R. Senger, H. Deogun, G. Carpenter, S. Ghiasi, T. Nguyen, N. James, M. Floyd, V. Pokala, A distributed critical-path timing monitor for a 65 nm high-performance microprocessor, in *IEEE International Solid-State Circuits Conference, 2007. ISSCC 2007. Digest of Technical Papers* (2007), pp. 398–399
13. J. Tschanz, K. Bowman, S. Walstra, M. Agostinelli, T. Karnik, V. De, Tunable replica circuits and adaptive voltage-frequency techniques for dynamic voltage, temperature, and aging variation tolerance, in *Symposium on VLSI Circuits, 2009* (2009), pp. 112–113
14. D. Ernst, N.S. Kim, S. Das, S. Pant, R. Rao, T. Pham, C. Ziesler, D. Blaauw, T. Austin, K. Flautner, T. Mudge, Razor: a low-power pipeline based on circuit-level timing speculation. In *Proceedings of the 36th Annual IEEE/ACM International Symposium on Microarchitecture, 2003. MICRO-36* (2003), pp. 7–18
15. M. Floyd, M. Allen-Ware, K. Rajamani, B. Brock, C. Lefurgy, A.J. Drake, L. Pesantez, T. Gloekler, J.A. Tierno, P. Bose, A. Buyuktosunoglu, Introducing the adaptive energy management features of the power7 chip. Micro, IEEE **31**(2), 60–75 (2011)
16. LEON3. http://www.gaisler.com/cms/
17. H. Cho, L. Leem, S Mitra, ERSA: error resilient system architecture for probabilistic applications. IEEE Trans. Comput. Aided Des. Integr. Circuits Syst. **31**(4), 546–558 (2012)
18. A. Rahimi, L. Benini, R.K. Gupta, Analysis of instruction-level vulnerability to dynamic voltage and temperature variations, in *Design, Automation Test in Europe Conference Exhibition (DATE)* (2012), pp. 1102–1105
19. AMD app SDK v2.5. http://www.amd.com/stream
20. X. Jiao, A. Rahimi, B. Narayanaswamy, H. Fatemi, J.P. de Gyvez, R.K. Gupta, Supervised learning based model for predicting variability-induced timing errors, in *2015 IEEE 13th International New Circuits and Systems Conference (NEWCAS)* (2015), pp. 1–4
21. X. Jiao, Y. Jiang, A. Rahimi, R.K. Gupta, Slot: a supervised learning model to predict dynamic timing errors of functional units, in *Design, Automation Test in Europe Conference Exhibition (DATE)* (2017)
22. X. Jiao, Y. Jiang, A. Rahimi, R.K. Gupta, Wild: a workload-based learning model to predict dynamic delay of functional units, in *2016 IEEE 34th International Conference on Computer Design (ICCD)* (2016), pp. 185–192

# Part II
# Detecting and Correcting Errors

As an alternative to methods in the previous part that seek to prevent timing errors from happening, in this part we examine methods that work through the timing errors [132, 137, 133, 135]. Presented methods here allow the timing errors to occur by operating at the edge of failure. Operating at the edge of failure further reduces the guardband. To combat the timing errors, these methods require two main mechanisms: (i) an error detection mechanism to detect the incorrect state values caused by the timing errors; and (ii) an error correction mechanism that is triggered upon an error detection to compensate the effects of errors during system operation. These methods span at different levels of software (Chap. 7) and architectures (Chap. 8).

The timing errors are typically corrected by error detection and recovery mechanisms at the circuit level. In contrast, in Chap. 7, we propose software methods for cost-effective countermeasures against hardware timing errors. Generally speaking, a shared-memory parallel architecture consists of a set of computing units that enables various options for executing a given workload. One of these options, as our central focus, is the selection of an appropriate computing unit to execute the workload. This choice between alternative computing units enables parsimonious execution of the workload in the presence of timing errors. Having such a choice, enables abstracting the errors from lower levels to higher levels that can lead to efficient error handling and better management. Accordingly, we implement a variability-aware OpenMP (VOMP) programming environment suitable for tightly-coupled sharedmemory processor clusters. VOMP is available as an extension to the OpenMP v3.0 programming model that covers various parallel constructs. Using the notion of work-unit tolerance as descriptive metadata, we capture timing errors caused by circuit-level variability as high-level software knowledge. As such, characterized metadata provide a useful abstraction of hardware variability to efficiently allocate a given work-unit to a suitable core for execution. VOMP enables hardware/software collaboration with online variability monitors in hardware and runtime scheduling in software providing 17% faster execution and 27% lower energy for embedded benchmarks parallelized with `task` directive. We further enhance proposed task scheduling strategies for simultaneous management of variability and workload by exploiting centralized and distributed approaches to workload distribution [133].

As another means to reduce the cost of error recovery, we focus on microarchitectures for computational reuse. Computational reuse, or memory-based computing, or memoization[1] refer to methods that normally use pre-computed results in place of actual computation at runtime. For instance, instruction reuse comes from the observation that many instructions can be skipped if another instance has already been executed using the same input values. The instruction reuse recalls the outcome of an instruction on a lookup table; therefore, a processor can reuse it temporally if the processor performs the same instruction with the same input values. Such lookup tables are typically implemented as an associative memory module. The associative memory modules can employ exact search (Chap. 8) or approximate search (Chap. 11) based on the accuracy requirements.

Parallel execution in GP-GPU architectures provides an important ability to reuse computation by associative memories and reduce the cost of recovery from timing errors. In Chap. 8, we devise an associative memory with exact search. This associative memory is coupled with a floating-point unit and recalls error-free operations to avoid the recovery cost in the event of timing errors. The associative memory exploits memristive nanodevices to improve density and cost of search operations. This technique enhances computational reuse using dense memristive nanodevices through monolithic 3D integration with CMOS at extremely low-cost.

[1]These three terms are used interchangeably through this book

# Chapter 7
# Work-Unit Tolerance

**Abstract** Manufacturing and environmental variations cause timing errors in microelectronic processors that are typically avoided by ultraconservative multi-corner design margins or corrected by error detection and recovery mechanisms at the circuit level. In contrast, we present in this chapter runtime software support for cost-effective countermeasures against hardware timing failures during system operation. We propose a variability-aware OpenMP (VOMP) programming environment, suitable for tightly coupled shared-memory processor clusters that relies upon modeling across the hardware/software interface. VOMP is implemented as an extension to the OpenMP v3.0 programming model that covers various parallel constructs, including `task`, `sections`, and `for`. Using the notion of work-unit vulnerability (WUV) proposed here, we capture timing errors caused by circuit-level variability as high-level software knowledge. WUV consists of descriptive *metadata* to characterize the impact of variability on different work-unit types running on various cores. As such, WUV provides a useful abstraction of hardware variability to efficiently allocate a given work-unit to a suitable core for execution. VOMP enables hardware/software collaboration with online variability monitors in hardware and runtime scheduling in software. The hardware provides online per-core characterization of WUV metadata. This metadata is made available by carefully placing key data structures in a shared L1 memory and is used by VOMP schedulers. Our results show that VOMP greatly reduces the cost of timing error recovery compared to the baseline schedulers of OpenMP, yielding speedup of 3–36% for tasks, and 26–49% for sections. Further, VOMP reaches energy saving of 2–46% and 15–50% for tasks and sections, respectively.

## 7.1 Introduction

The most immediate manifestation of variability is in path delay variations. Path delay variations cause violation of timing specification resulting in circuit-level timing errors. Timing errors can result in an *errant* instruction leading to a malfunction within the computing core. Hence, robust system design needs to ensure that systems perform correctly despite increasing timing failures caused by variability in



A. Rahimi et al., *From Variability Tolerance to Approximate Computing in Parallel Integrated Architectures and Accelerators*,
DOI 10.1007/978-3-319-53768-9_7

many-core processor chips [1]. To ensure correct functionality in the presence of timing error, some approaches rely upon error recovery mechanism that guarantees correct program execution eventually. The timing failures are typically corrected by either adaptive tuning of CMOS control knobs to provide better-than-worst case guardband for error-free instruction execution [2], or by *replaying* the errant instruction [3]. For instance, a 45 nm Intel resilient core [3] places EDS sensors [4] at the endpoints of the critical paths of the pipeline stages. Once a timing error is detected during instruction execution, the core prevents the errant instruction from corrupting the architectural state and an error control unit (ECU) triggers proper actions to ensure error recovery. The ECU first flushes the pipeline to resolve any complex bypass register issues, and then triggers one of the two recovery mechanisms: (1) instruction replay at half clock frequency; (2) multiple-issue instruction replay at the same clock frequency. These mechanisms impose energy overhead and latency penalty of up to 28 extra recovery cycles per error [3] which can adversely affect both performance and energy [5].

To achieve the required robustness while reducing these overheads, the variability-induced timing errors can be addressed through a combined hardware–software approach [6–8] that allows to evaluate the impact of a timing error on the overall system. We have shown in the previous chapters that a holistic cross-layer variability management can abstract the circuit-level timing error information into the vulnerability of individual (Chap. 2) or streams (Chap. 3) of instructions when executed on a particular core. For multi-core processors, this knowledge can be used by the runtime system to implement variability-tolerant parallel workload deployment for reducing the cost of timing error failure correction [9]. We have earlier defined a set of hierarchically organized vulnerability measures—from instruction set architecture to a parallel programming model—to expose variations and their effects to the software stack. These measures include instruction-level vulnerability (ILV) [10], sequence-level vulnerability (SLV) [11], procedure-level vulnerability (PLV) [12], and finally task-level vulnerability (TLV) [9]. ILV characterizes individual instructions as the most fine-grained abstraction of the processor's functionality, while SLV determines streams of instructions that have a significant impact on the timing error rate. Raising further the level of abstraction, PLV exposes the effect of dynamic voltage variations for use in software preventive actions. Within a shared-memory multi-core computing cluster, PLV enables a runtime procedure hopping technique to mitigate the effect of variations by means of low-cost subroutine (procedure) migration to a less vulnerable core [12]. TLV is an extension to the OpenMP v3.0 tasking programming model to dynamically characterize the vulnerability of tasks. Here, the runtime system reduces the cost of error recovery by matching the characteristics of different variability-affected cores to the vulnerability of individual parallel tasks.

In this chapter, we extend the definition of TLV to that of work-unit vulnerability (WUV), where the notion of a parallel work-unit (WU) is specialized into any of three OpenMP constructs to specify work-sharing among parallel threads: `task`, `sections`, and `for`. Our goal is to provide runtime software support to increase cost-effective countermeasures against timing errors in hardware. We pursue this goal by exposing variability and its effect to the OpenMP programming model, thus enabling holistic variability management. Accordingly, we make three contributions:

1. We devise a variation-aware synergistic hardware/software approach. It enhances robustness of cluster-based processors through cost-effective software countermeasures against timing failures in hardware during system operation. On the hardware side, our multi-core cluster is equipped with circuit sensors for online measurement of variability and per-core introspective *metadata* characterization for a given workload. Fast access to metadata for each type of OpenMP work-sharing construct is guaranteed by carefully placing the key data structures on fast shared-L1 memory.
2. On the software side, we propose a fully variation-aware OpenMP (VOMP) environment, which supports `task`, `sections`, and `for`. VOMP provides online characterization of descriptive metadata for these constructs. Characterized WUV, or work-unit vulnerability, abstracts hardware variability that reflects the manifestation of circuit-level timing errors during the execution of an instance of a specific OpenMP construct. We also propose a set of scheduling algorithms that implement software-only countermeasure schemes, one for each work-sharing construct. Hence, the OpenMP runtime scheduler utilizes WUV metadata during scheduling to efficiently mitigate the variability-induced timing errors at the level `tasks` and `sections`. This leads to a holistic runtime management system that strives to reduce the cost of error recovery caused by execution of various work-sharing constructs.
3. We demonstrate the effectiveness of our approach on a variability-affected tightly coupled processor cluster with accurate ILV models in 45-nm TSMC technology. Our experimental results indicate that (i) the entire cost of online software characterization and countermeasures is paid off for a variability-affected fabric. (ii) The proposed VOMP environment is able to save both energy and total execution time for a wide range of parallelized applications. VOMP reduces the execution time by 3–36% and energy by 2–46% for applications parallelized with `task` directives. VOMP also reaches to energy saving of 15–50% and faster execution of 26–49% for applications using `sections` directives. Further, we evaluate the robustness of our approach across 80 °C temperature variations.

The rest of this chapter is organized as follows. Section 7.2 covers the architectural details to support VOMP. Section 7.3 describes characterization of WUV metadata for every type of work-unit under a full range of dynamic voltage ($\Delta V = 0.22$ V) and temperature ($\Delta T = 140$ °C) variations. The proposed runtime scheduling algorithms for each work-sharing construct are presented in Sect. 7.4. In Sect. 7.5, we explain our methodology to capture variations, framework setup, and present experimental results followed by conclusions in Sect. 7.6.

## 7.2 Architectural Support for VOMP

We now describe the architectural details of the variation-tolerant processing cluster, shown in Fig. 7.1. The architecture is inspired by STMicroelectronics Platform 2012 (P2012) [1, 13] as a programmable many-core accelerator for next-generation data-intensive embedded applications. The P2012 computing fabric is modular and scalable, since it is based on multiple processor clusters such as those found in GP-GPUs [14] and clustered accelerators like HyperCore architecture line processors from Plurality [15], and Kalray multipurpose processor array [16]. Every cluster has independent power and clock domain, therefore enabling fine-grained power and variability management [1]. The clusters are connected via a fully asynchronous network-on-chip that enables them to work with different clock frequencies decided by a cluster controller for the power/variability management [1]. In our implementation, we focus on a single cluster consisting of s16 tightly coupled 32-bit in-order RISC cores, a level-one (L1) tightly coupled data memory (TCDM) and a low-latency 16×32 logarithmic interconnection [17]. The TCDM is a software-managed scratchpad memory, configured as a shared, multi-ported, multibanked L1 memory that is directly connected to the logarithmic interconnection for fast accesses. The number of TCDM ports is equal to the number of banks (32) to enable concurrent access to different memory locations. Note that a range of addresses mapped on the TCDM space provides test-and-set read operations, which we use to implement basic synchronization primitives, e.g., locks.

The logarithmic interconnection is composed of mesh-of-trees networks to support *single cycle* communication between the cores and TCDM banks (see the left part of Fig. 7.1). When a read/write request is brought to the memory interface, the data is available on the negative edge of the same clock cycle, leading to two clock cycles latency for a conflict-free TCDM access. The cores have direct access into the off-cluster L2 memory, also mapped in the global address space. Transactions to the L2 are routed to a logarithmic *peripheral interconnect* through a demultiplexer stage. From there, they are conveyed to the L2 via the system interconnection which is based on the AHB bus. Since the TCDM has a small size (256KB) the software

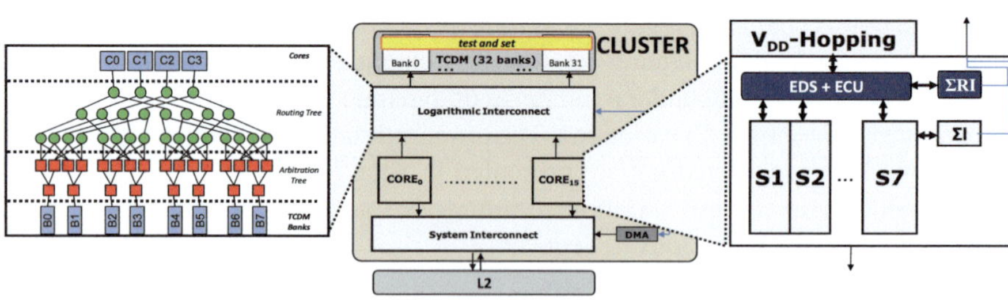


**Fig. 7.1** Variation-tolerant tightly coupled processor cluster for VOMP. The *right* part shows a resilient core with EDS and ECU to correct timing errors by the replica instructions; $\Sigma I$ is the number of error-free instructions, and $\Sigma RI$ is the number of replayed instructions

must explicitly orchestrate continuous data transfers from L2 to L1, to ensure locality of computation. To allow for performance- and energy- efficient transfers, the cluster has a DMA engine. This can be controlled via memory-mapped registers, accessible through the peripheral interconnect.

In the embedded tightly coupled processor cluster, it is essential that all the cores within a cluster work with the same clock frequency to avoid the latency of the synchronization [1]. Synchronization across multiple frequencies increases the latency of the interconnection, and has a performance penalty as high as a L1 cache miss[1] [17]. Therefore, the cores within the cluster are equipped with two circuit-level resiliency techniques. First, each core relies on the EDS [4] circuit sensors to detect any timing error due to dynamic delay variation. To recover the errant instruction without changing the clock frequency, the core employs the multiple-issue instruction replay mechanism [3] in its error recovery unit (ECU). It issues seven replica instructions (equal to the number of pipeline stages) followed by a valid instruction. Second, the cluster supports a $V_{DD}$-hopping technique [18] that discretely tunes the voltage of slow cores—the cores that are affected by static process variation. The $V_{DD}$-hopping improves the clock speed of the slow cores, thus enabling all the components of the variability-affected cluster to work at same frequency (with memories at a 180° phase shift). This technique avoids the inter-core synchronization that would significantly increase L1 TCDM latency. The core-level $V_{DD}$-hopping has been already employed in a variability-tolerant tightly coupled cluster [12]. However, a core with higher vulnerability will impose extra cycles to correct the errant instructions.

## 7.3 Work-Unit Vulnerability and VOMP Work-Sharing

OpenMP [19] consists of a set of compiler directives and library routines to specify parallel execution within a sequential code. Enclosing a code block within a `#pragma omp parallel` directive has the effect of launching multiple instances of that code over the available processors. Differentiating the actual work done by different processors in OpenMP is achieved by means of work-sharing constructs: `#pragma omp for`, `#pragma omp sections`, and `#pragma omp task`. The `for` directive can only be associated to a loop nest, and distributes loop iterations over available processors. Within a `sections` directive multiple section blocks can be specified, each containing a different parallel work-unit. Sections have limited expressiveness for describing task parallelism. For this reason, the latest OpenMP specifications have included the new `task` directive, which supports sophisticated forms of task parallelism. However, `task` implies significant overheads, which makes `sections` more convenient to outline few coarse-grained tasks in a program. In addition, it is easy to describe software pipeline parallelism

[1] 8 cycles are required for synchronization between multiple clock domains for a read/write operation, while performance of the architecture relies on the fact that we have 2 cycles access to L1 memory.

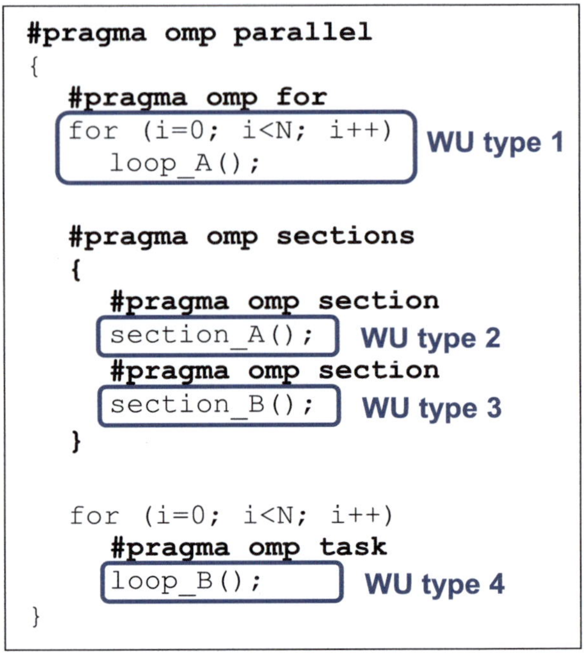


**Fig. 7.2** Outlined WU types in a OpenMP program: `task`, `sections`, `for`

with `sections`, by just adding point-to-point synchronization to enforce dependencies within parallel tasks. The latter is the main use we make of `sections` in this chapter.

As discussed earlier in the introduction, to enable software-driven policies for variability-tolerant parallel workload scheduling we need to characterize parallel work-units, WU, in terms of vulnerability to timing errors.[2] Each OpenMP work-sharing construct outlines an execution unit which runs a sequence of instructions. Enclosing portions of code within any of these constructs allows the programmer to statically identify several WU *types* in the program, as every directive syntactically delimits a unique stream of instructions. While at runtime the same stream may be dynamically instantiated several times (e.g., a work-sharing directive nested within a loop), from the point of view of our characterization it uniquely identifies a single WU type. As a direct consequence, there are as many types of WUs in a program as there are work-sharing directives in its code, as shown in Fig. 7.2.

Intuitively, the closer we can associate information on variability-induced timing errors (*metadata*) to software abstractions of a parallel WU, the better we can schedule WUs to cores in a variation-tolerant manner. From this perspective, task-level vulnerability, or TLV, is an important metadata to address variability tolerance within standard parallel programming models. The main limitation of TLV as described in [9] is that its implementation is specific to the `task` OpenMP construct. While

[2]Our platform does not have control over the errors happening while executing library code. The functionality is preserved as each core is equipped with the replay mechanism.

this construct allows to express very flexible and sophisticated forms of dynamic parallelism, it is also true that several embedded workloads focus on more regular forms of parallelism, at the loop- or procedure-level [20]. Until the specification v2.5 OpenMP used to be focused on exactly those types of parallelism, through the `for` and `sections` constructs.

In our previous work [10] we have introduced ILV or instruction-level vulnerability as a metric to expose to the software stack the effect of variations on the performance of a processing core, at the level of individual instructions. In a variability-affected core ILV is not uniform across the instruction set. In fact, ILV partitions instructions into three classes: (i) logical/arithmetic, (ii) memory, (iii) hardware multiply/divide. Instructions belonging to different classes have different vulnerability to variations depending on the way they exercise the non-uniform critical paths across the various pipeline stages. For instance, in an in-order RISC core the execution and memory stages are highly vulnerable to dynamic variations, and the memory class has a higher vulnerability in comparison to the logical/arithmetic class. We note that complex out-of-order core such as IBM POWER6 also confirms that vulnerability is not uniform across the instructions set [21].

Here we extend the notion of ILV to a more coarse-grained (in terms of software execution units) metric: parallel *work-unit vulnerability* (WUV). WUV is a metric to estimate execution time of each WU type per each core, under variability. This metric is quite useful for the purpose of simultaneous vulnerability measurement and load balancing. The vulnerability of a WU type varies based on the class of instructions that it executes. WUV is clearly a per-core metric since the amount of variation affecting different classes of instructions changes from one core to another. Therefore, different dynamic instances of the same WU type can face different degrees of variability-induced timing errors.

While the identification of WU types can be done statically (i.e., at compile time), WUV characterization has to be done online due to two main reasons. First, dynamic instances of the same WU type may exercise the processor pipeline in a non-identical manner due to data-dependent control flow that results in the execution of different (classes of) instructions. Second, the characterization must reflect the variability-affected characteristic of every core (not known a priori) on every WU type. WUV is defined as follows:

$$WUV_{(i,j)} = \sum I + \sum RI \mid \forall core_i, \forall WUtype_j, \tag{7.1}$$

where $\Sigma I$ is the number of error-free executed instructions; $\Sigma RI$ is the number of replayed instructions[3] during execution of WU type $j$ on core $i$, as reported by the ECU. Intuitively, for a given WU type if all the instructions run without any timing error, the corresponding WUV is equal to $\Sigma I$ as the total error-free dynamic instruction count. In the event of timing errors, WUV also accounts for the additional replica instructions. The lower the WUV, the lower number of recovery cycles, the lower the dynamic instruction count, and thus the higher throughput and energy

[3]Proportional to the number of errant instructions.

efficiency. WUV dynamically characterizes both vulnerability and execution time of WU types. Hence based on WUV values, VOMP runtime schedulers can optimize the system performance or energy efficiency by matching variability-affected core characteristics to WU types.

### 7.3.1 Intra- and Inter-corner WUV

For Eq. 7.1 WUV is the dynamic instruction count, including the replica instructions, for a given WU type. Similar to ILV, WUV is also not uniform across different variability-affected cores, which may exhibit different vulnerability to specific instruction classes. To demonstrate how this effect is propagated to the programming model level, we measure WUV across different WU types. More specifically, we use OpenMP constructs to outline software execution units, or WUs, which iterate several times over an identical instruction. We build four WU types each stressing a different instruction, as shown in Fig. 7.3.

In the following, we repeat the same experiment with different OpenMP work-sharing constructs. This synthetic experiment allows to stress a use case where we can estimate the variations in WUV among the software execution units. Figure 7.4 illustrates the synthetic benchmark parallelized with the `#pragma omp task` construct, while the synthetic benchmark in Fig. 7.5 uses the `#pragma omp`

**Fig. 7.3** WU types each stressing a different class of instructions

```
#define OP_MUL   1
#define OP_ADD   2
#define OP_DIV   3
#define OP_SHIFT 4
int A[][][], B[][][], C[][][];
void WU_run (int z, int OP)
{
  for (int y = 0; y < N; y++)
    for (int x = 0; x < N; x++)
    {
      switch(OP)
      {
        case OP_MUL:   C[x][y][z] =
            A[x][y][z] * B[x][y][z];
            break;

        case OP_ADD:   C[x][y][z] =
            A[x][y][z] + B[x][y][z];
            break;

        case OP_DIV:   C[x][y][z] =
            A[x][y][z] / B[x][y][z];
            break;

        case OP_SHIFT: C[x][y][z] =
            A[x][y][z] >> B[x][y][z];
            break;
      }
    }
}
```

Fig. 7.4 Synthetic benchmark using OpenMP `task`

```
#pragma omp parallel
{
  #pragma omp master
  {
    for (int z = 0; z < N; z++)
      #pragma omp task
        WU_run (z, OP_MUL);

    for (int z = 0; z < N; z++)
      #pragma omp task
        WU_run (z, OP_ADD);

    for (int z = 0; z < N; z++)
      #pragma omp task
        WU_run (z, OP_DIV);

    for (int z = 0; z < N; z++)
      #pragma omp task
        WU_run (z, OP_SHIFT);
  }
}
```

Fig. 7.5 Software pipelined synthetic benchmark using OpenMP `sections`

```
#pragma omp parallel
{
  for (int z = 0; z < N; z++)
  {
    #pragma omp sections nowait
    {
      #pragma omp section
      {
        WU_run(z, OP_MUL);
        synch();
      }
      #pragma omp section {
        synch();
        WU_run(z, OP_ADD);
        synch();
      }
      #pragma omp section {
        synch();
        WU_run(z, OP_DIV);
        synch();
      }
      #pragma omp section {
        synch();
        WU_run(z, OP_SHIFT);
      }
    }
  }
}
```

`sections` construct. For the sake of clarity, we organize the presentation of this experiment in following three consecutive subsections, one per each OpenMP construct. Section 7.5.1 provides details of our simulation setup.

#### 7.3.1.1 `task`-Level WUV

Figure 7.4 shows the synthetic benchmark parallelized using the `#pragma omp task` construct. We measure WUV for different WU (here, `task`) types, when executing on fixed and variable operating corners (current voltage and temperature). Specifically, we analyze the effects of a full range of operating corners, a temperature range of 0–140 °C, and a voltage range of 0.88–1.1 V. For sake of simplicity, in this section we illustrate a *normalized* WUV (thereafter called NWUV) as a metric which divides WUV value to its $\Sigma I$, therefore this normalized metric will have a range of values greater than or equal to 1. For instance, if NWUV displays a value of 1, it indicates that there is no replica instructions ($\Sigma RI$=0).

Figure 7.6 shows the task-level WUV for a core that works at fixed voltage supply of 1.1 V, while the environmental temperature is varied. As shown, the `task`-level vulnerability is an increasing function of temperature; for instance, the execution of task type one ($task_1$) at a temperature of 0 °C results in an NWUV value of 1.0017, while executing the same task at 140 °C causes an NWUV of 1.09 that

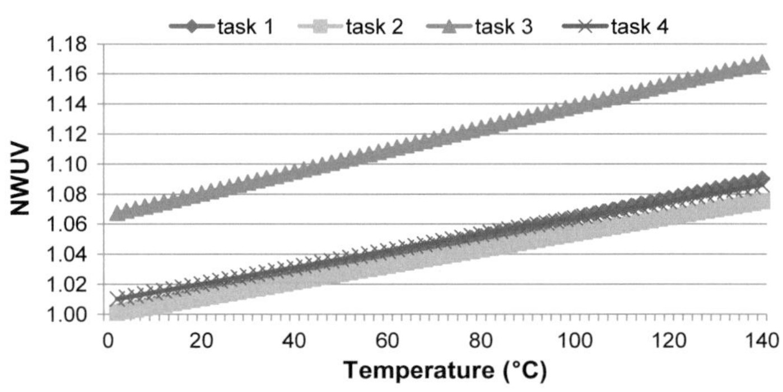


**Fig. 7.6** Normalized WUV (NWUV) to temperature variations for `task` types

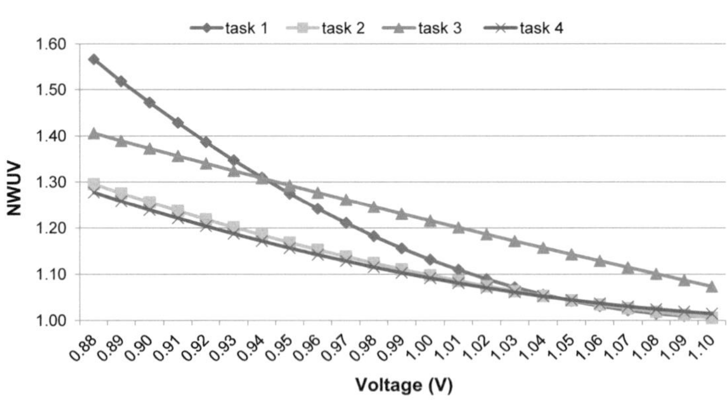


**Fig. 7.7** Normalized WUV to voltage variations for `task` types

increases the vulnerability of $task_1$ by 9%. This inter-corner WUV variation is the direct manifestation of dynamic temperature fluctuation. At supply voltage of 1.1 V, higher temperature leads to a higher timing error rate that increases the number of errant instructions, as mirrored by the WUV values.

Apart from the inter-corner WUV variation, for a given (fixed) temperature point there is an intra-corner WUV variation among the four types of WUs (tasks). As shown in Fig. 7.6, at the fixed temperature of 0 °C, the WUV value of $task_3$ is 6% higher than the WUV of $task_2$, indicating a considerable variation across task types. WUV of each task type is different, even within the fixed operating conditions and in the absence of environmental variations, since each task type executes distinct classes of instructions experiencing different rates of the errant instructions.

Figure 7.7 shows the `task`-level WUV for the core operating at a fixed temperature of 10 °C, while voltage is dynamically varied. As shown by the plot, NWUV is a decreasing function of voltage. Higher voltages result in shorter critical path delay, thus lower error rate and finally lower NWUV values. Similar to Fig. 7.6, intra-corner WUV variation can also be observed: WUV for different task types at the same operating corner is not equal because their instructions do not uniformly exercise the various critical paths of the pipeline. We have already seen that the vulnerability of instructions is not uniform [10] resulting in different levels of vulnerability for task types.

#### 7.3.1.2 `sections`-Level WUV

Figure 7.5 shows the code for the synthetic software pipeline implemented using parallel `sections`. Each WU type (here indicated as $section_1$, $section_2$, $section_3$ and $section_4$) is mapped on a different core. Synchronization between the pipeline stages is accomplished via simple point-to-point synchronization primitives that we implement on top of test-and-set semaphores. This guarantees that once computation of one pipeline stage is finished we can start the following stages. The `sections` construct is nested within a loop, which models the repetitions of the pipeline. It outlines four WUs, each dependent from the previous one. Note however that there is no dependence between the last stage of one iteration and the first stage of the next iteration.

In this parallel pattern, representative of image processing kernels where a set of filters is applied in sequence to independent image blocks (e.g., JPEG macro-blocks), there are $N_{sec}$ stages, such that $N_{sec} < N_{core}$, where $N_{core}$ is the number of available cores (16 cores in our platform). Normally, at the end of any work-sharing construct it is implied a barrier synchronization operation among all processors. However, we specify the `nowait` clause to skip this and allow the idle cores to start execution of the next pipeline iteration.

We now examine the `sections`-level WUV for different section types when executing on fixed and variable operating corners. Figure 7.8 shows NWUV values for a core operating at fixed supply voltage of 1.1 V with a variable temperature range of 0–140 °C, while Fig. 7.9 shows NWUV values for a fixed temperature of 10 °C

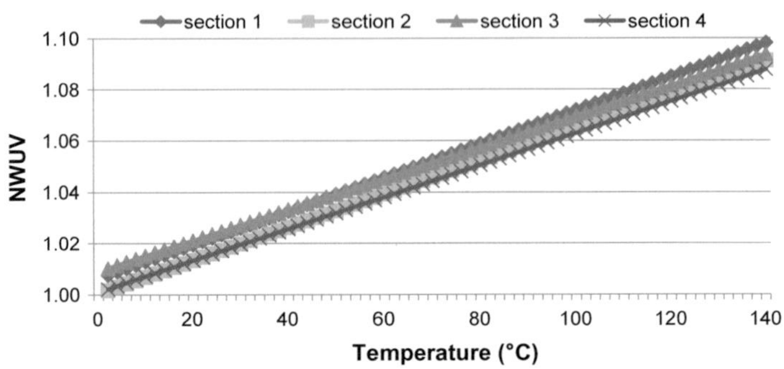


**Fig. 7.8** Normalized WUV to temperature variations for `sections` types

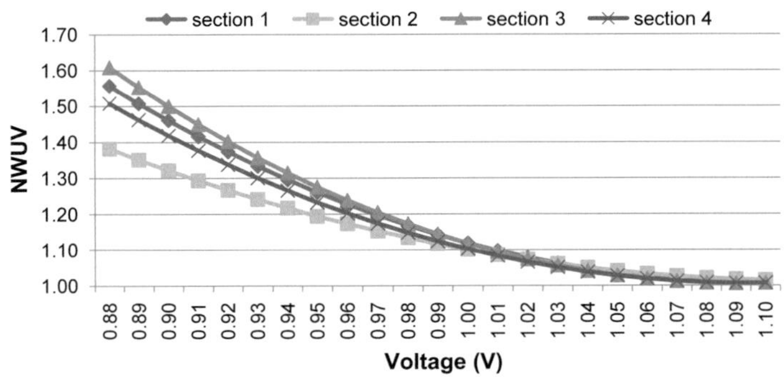


**Fig. 7.9** Normalized WUV to voltage variations for `sections` types

with a supply voltage variation range of 0.22 V. Akin to the `task`-level WUV, the `sections`-level WUV is an increasing function of temperature and a decreasing function of voltage. A temperature fluctuation of 140 °C increases the `sections`-level WUV by an average of 9%, and the voltage variation of 0.22 V increases the `sections`-level WUV by an average of 50%. Among the different section types, a maximum of 16% intra-corner WUV variation is observed at (10 °C, 1.09 V).

#### 7.3.1.3 `for`-Level WUV

Applications running on multi-core systems often focus on a very common data parallel scenario where each core works on a portion of a data structure (e.g., array or matrix) and must synchronize with the others on a barrier. Similar parallelization schemes are typically focused on parallel loops, whose iterations are spread among several concurrent threads. Data-level parallelism, for instance parallel loops, can be exploited to distribute workload within a cluster. OpenMP v3.0 provides dynamic loop scheduling as another work-sharing construct based on the notion of a work queue to parallelize loops locally inside a cluster. A `parallel for` directive describes a loop as a set of identical work-units; therefore the `parallel for` directive statically identifies one type of work-unit in the program. For every loop iteration, the work-unit is dynamically instantiated but it uniquely identifies a single type from our characterization point of view. In other words, the work-units generated from the `parallel for` directive are equivalent hence forming a *homogeneous* workload across all cores. This limits the capability of VOMP to schedule a single type work-unit to an appropriate core given that maintaining all cores busy.

#### 7.3.1.4 Conclusion for WUV

The main conclusion that we can draw from the experiments presented in the aforementioned subsections is that WUV varies significantly: (i) among WU types; and (ii) among the operating conditions. On one hand, this is due to how different instruction streams exercise the variability-affected critical paths in the processor pipelines, which is the typical case for programs parallelized with `sections` or `task` directives that outline several parallel tasks (i.e., WU types). This confirms the previous observation that executing different streams of instructions may result in various error rates [22]. For example, for any given operating condition the WUV of simple arithmetic operations (e.g., addition/shift) is lower than or equal to the WUV of complex arithmetic operations (e.g., MUL/DIV). Detailed sensitivity analysis of a sequence of instructions to changes in voltage and temperature is provided in [11]. On the other hand, even identical instruction streams behave differently on different cores in presence of dynamic temperature and voltage variations. This is particularly evident for the `#pragma omp for` construct, which always distributes among processors an identical work-unit type (i.e., the same instruction stream). Yet, WUV across cores varies significantly, because of the different vulnerability to specific instruction classes and to operating conditions.

This motivates the need to specialize WUV for different WU types and for online characterization. In the following section, we describe how we augment the VOPM runtime support for each of the work-sharing constructs to support online WUV characterization.

### 7.3.2 *Online WUV Characterization*

In the proposed VOMP, each core performs online characterization while executing a given WU type. To quantify WUV, the core collects $\Sigma I$ and $\Sigma RI$ statistics for Eq. 7.1 through a set of available counters in the ECU. The online characterization mechanism is distributed among all the cores in the cluster, thus enables full parallel WU execution monitoring and characterization. WUV is represented as a two-dimensional lookup table (LUT) for different WU types and cores. This lookup table is physically distributed across all the banks of the L1 TCDM for fast parallel read/write operations. Since each entry of the LUT consists of 32-bit integer data, and since each application includes a bounded[4] number ($N_{WU}$) of work-sharing directives, the LUT has a footprint of $N_{WU} \times 4 \times N_{core}$ Bytes, $N_{core}$ being the number of cores in the cluster. We provide two simple functions for reading and writing the LUT, namely:

```
int LUT_rd (int WUtype, int coreID);
void  LUT_wr (int WUtype, int coreID,
    int WUV);
```

In addition, we implement two functions for retrieving the calculated WUV of a task running on a core.

```
int read_WUV (int coreID);
void    reset_WUV (int coreID);
```

The former function (`read_WUV`) reads the WUV value from per-core hardware counters, identified via the `coreID` parameter. These counters implement Eq. 7.1, accumulating instruction count and replica instruction count for the target core since the last reset. The second function (`reset_WUV`) resets the counter for the target core (`coreID`).

Based on these low-level APIs, we modify the OpenMP runtime schedulers to enable online WUV characterization as illustrated in Fig. 7.10 (our additions in **bold font**). While this pseudocode explicitly refers to the `task` scheduler, we modify in an equivalent manner also the scheduler for `sections`. For what concerns loops the implementation is slightly more complicated. OpenMP allows to couple the `schedule(static|dynamic)` clause to the `#pragma omp for` directive. Choosing dynamic scheduling, chunks of iterations of user-defined size are scheduled to parallel cores in a first-come, first-served manner. This allows for better load balancing at runtime, but is implemented through calls to a runtime scheduler and implies higher overhead. For those cases where loop iterations contain identical amount of works it is often better to use static scheduling, which is implemented by statically inlining the code that precomputes the assigned iterations to any cores. Thus, for dynamic scheduling we instrument the runtime scheduler similar to Fig. 7.10. For static scheduling, we modify the OpenMP compiler to inline the additional WUV characterization code during the loop expansion pass.

Note that in principle, it would be strictly necessary to characterize a couple <*WUtype*, *coreID*> only once. Once a WU type is characterized for a given core the online characterization could be stopped. However, we rather keep the characterization active at every scheduling event and apply a history-based weighted average calculation between the new characterized WUV value and the previously WUV value stored in the LUT. This has been used to estimate power and time for a given interval [23]; and also better captures recent effects of dynamic variations on the cores, conditional code within WUs, and future workload. At each scheduling point,

[4]Up to a few tens, for large programs.

Fig. 7.10 Pseudocode for task-level WUV characterization

```
When task_j is scheduled on core_i:
begin
  EXTRACT_TASK (task_j)
  WUV_old = LUT_rd (task_j, core_i)
  reset_WUV (core_i)
  EXECUTE_TASK (task_j)
  WUV_new = read_WUV (core_i)
  WUV_write = (WUV_new-(WUV_new>>3))+(WUV_old>>3)
  LUT_wr (task_j, core_i, WUV_write)
end
```

the encountering core incurs only a fixed negligible overhead for WU characterization. This is achieved by distributing the LUT in the multibanked TCDM that enables not only *predictable* accesses, as opposed to cache-based hierarchical memories, but also fast parallel read/write operations among the cores.

From the observation point of view, our online characterization can reflect any changes in dynamic behavior of a core and the environment in which the core is used. More specifically, in our cluster each core can be powered at a different voltage (that could lead to different temperature points due to self-heating), but all the 16 cores have to work with a fixed clock frequency. Figures 7.6, 7.7, 7.8, and 7.9 show the sensitivity of WUV to changes in the operating voltage and temperature. These figures illustrate that a wide range of dynamic variations can be reflected by WUV metric. From the controllability point of view, the cluster as an accelerator operate under the control of a main *host* processor, capable of running full-fledged operating systems (OS). The cluster itself, on the other hand, typically does not have all the necessary support to run unmodified OS. Resource management is demanded to custom lightweight middleware. In this respect, the OpenMP implementation that we leverage in this work [24] as a baseline to demonstrate our techniques is designed to operate on *bare metal*, as it is built directly on top of the hardware abstraction layer (HAL). The HAL provides the lowest level software services for processor (thread) and memory management, as well as the power control APIs.

## 7.4 VOMP Schedulers

### *7.4.1 Variation-Aware Task Scheduling (VATS)*

In this subsection, we first explain our OpenMP tasking implementation followed by our specific variation-aware scheduling policy. OpenMP tasking has already been considered as a convenient programming abstraction for embedded multi- and many cores [9, 25–27]. Typically in these approaches, the task scheduler is implemented using a centralized queue which collects the task descriptors. The central FIFO design reduces the overhead for task management, which is usually a relevant design choice

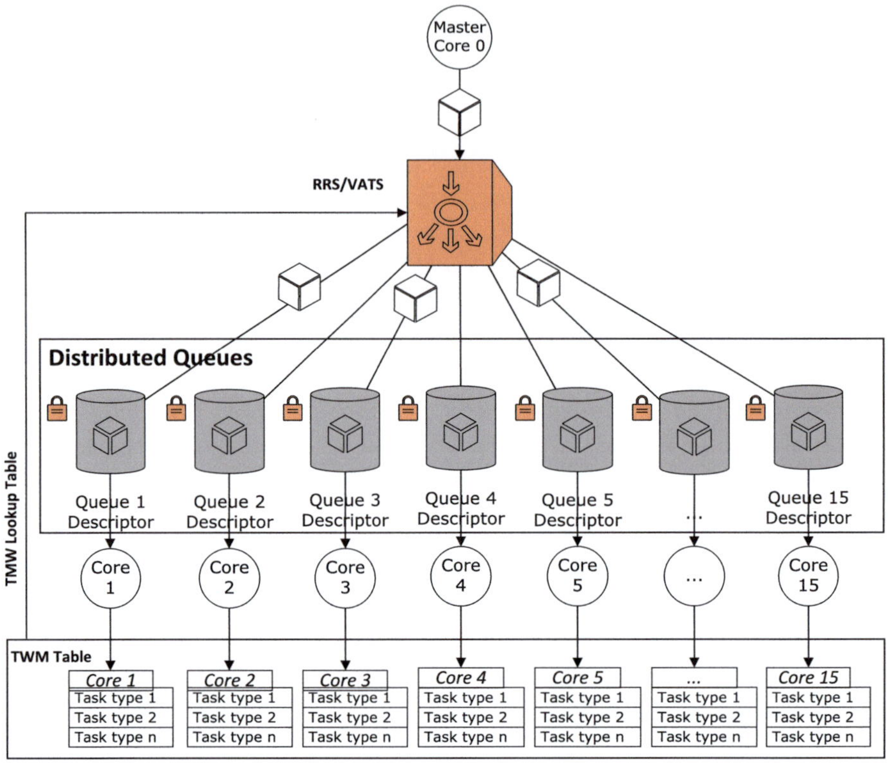


**Fig. 7.11** Distributed queues for OpenMP tasking

for energy- and resource-constrained systems. This design choice works well for homogeneous systems, but places limitations on applying efficient scheduling policies in presence of variability-induced heterogeneity across computational resources.

Our OpenMP implementation leverages distributed task queues (private queue per each core), where all the threads[5] involved in parallel computation can actively push and pop job descriptors. Figure 7.11 shows the design of our OpenMP tasking framework based on a distributed queue system. Every thread can access a queue using two basic operations: *insert* and *extract*, which are translated into lock-protected operations on a queue descriptor (stored in TCDM for minimal access time). Queue descriptors are statically instantiated during the initialization of the runtime to avoid the time overheads for dynamic memory management. Since threads with an empty queue are set to a low-power IDLE mode, the insertion of a task in a queue wakes up the associated core. This is achieved by inspecting an additional flag of the queue descriptor, where the destination core operating mode is annotated (*executing*, *sleeping*). The core that inserts the task in a remote queue is responsible for checking the

[5]There is a 1:1 correspondence between threads and cores, thus we will use the two terms interchangeably.

flag and waking up the destination core to resume execution of the newly inserted task. In addition, the queue descriptor holds synchronization flags used for the `taskwait` directive. Extracting a task from a queue updates the queue descriptor in the dual manner. Note that also in this case we use lock-protected operations, since we allow all threads to extract work from any queue. Extracting tasks always occurs from the head of the queue, while insertion can be done at the head and tail. Insert operations at the head are useful to prioritize the execution of non-characterized tasks (in terms of vulnerability to errors). Stealing tasks occurs from the head of the queue.

As a baseline policy we implement a simple round-robin scheduler (RRS) [19]. This policy aims at balancing the number of tasks assigned among all cores, and introduces minimal runtime overhead due to a very lightweight implementation. To account for tasks of different durations, RRS is enhanced with a task stealing algorithm, which searches remote queues in a round-robin fashion for work to steal.

We propose a reactive policy for variability-aware task scheduling (VATS) shown in Algorithm 7.4.1. This scheduler leverages the characterized WUV metadata to allocate tasks to cores so as to minimize both overall number of instruction replays and unbalanced loads. The main goal of this scheduler is to prevent allocation of tasks to unreliable cores, which is representative of a policy adopted in a system where task failure has critical consequences. At system startup, when there is no WUV available, the scheduler operates in round-robin mode. Since the OpenMP tasking model assumes completely independent tasks, it is allowed to execute them in any order. We leverage this property to insert tasks for which WUV is not available yet at the head of the queue (*out-of-order* task characterization). This will give higher priority to non-characterized task types, thus speeding up the "system warm-up".

---

**Algorithm 7.4.1.** VATS $(task_j)$

---

**for** $i \leftarrow 1$ **to** $N_{core}$

**do** $\begin{cases} load_i \leftarrow loadQueue_i + WUV(core_i, task_j) \\ min \leftarrow findMinimum(load_i) \end{cases}$

$Queue_{min} \leftarrow insert(task_j)$

**return** $(min)$

---

VATS scheduling policy strives to minimize the number of replayed instructions utilizing characterized WUV metadata. VATS also extends its awareness of the load on each queue, thus avoids heavily unbalanced situations that could increase the total execution time. Each queue descriptor is enhanced with a status register that estimates the overall load (*loadQueue*), in terms of dynamic instructions count, of all tasks present into that queue. This is a better metric for workload awareness than just the total task count, because different task types present in the queue may have various computational weight.

To account for imbalance effects due to non-homogeneous task durations and other system-level issues, VATS is further enhanced with a *most loaded queue-first* stealing algorithm. An additional array structure is used to keep the sorted workload over the various queues. This array is then traversed to steal work from the most loaded queues first. Note that after the execution of a stolen task, we always check if in the meantime some tasks have been inserted in the local queue. It this case, we switch to the execution of the tasks with better WUV values, otherwise we continue executing the stealing algorithm until there is no task left in the system.

### *7.4.2 Variation-Aware Section Scheduling (VASS)*

The default OpenMP section scheduling policy is to allocate a section to an available thread in a first-come, first-served (FCFS) fashion. When sections are used in a traditional manner to outline parallel tasks with no dependencies among each other Algorithm 7.4.1 can be applied. However, when sections are used to model software pipeline parallelism we have an additional constraint: avoiding the variability-induced errors (hence their instruction replays) that lengthen in an uncontrolled manner one or more sections. This effect dominates the overall pipeline duration. Since in a variability-affected computing cluster, there might be a set of cores that display poor performance—depending upon their software and hardware context—causing bottlenecks in the entire pipeline execution.

For these cases, we propose a variation-aware section scheduling (VASS) policy shown in Algorithm 7.4.2. VASS has a *warm-up* phase which assigns execution of different section types to all cores for a constant[6] number of iterations. After execution of each section, the characterization process updates the corresponding WUV metadata in LUT using the mechanisms described in Sect. 7.3.2. When the warm-up phase is completed, the WUV metadata in the LUT are ready and can be inspected by the runtime environment to take decisions on workload distribution. Accordingly, VASS assigns the execution of each section to a set of suitable cores.

In this way, VASS strives to maintain all cores in the *executing* operating mode, while reducing the instruction replays and the overall pipeline duration. VASS sorts each section types based on their average WUV decreasingly. The first section type in the sorted list has either high instruction count ($\Sigma I$) or high replica instruction count ($\Sigma RI$). Therefore it should be executed on a set of suitable cores that display fewer error rate during its execution. Basically, every core has a private tag vector that lists the types of *permissible* sections for executing on this particular core. This constraint limits the participation of worse cores for executing long or high vulnerable types of sections. The worse cores instead may execute shorter sections or sections with lower vulnerability; therefore avoiding the latency penalty for the synchronization between the unbalanced stages and effectively utilizing all the resources in the variability-affected cluster.

As shown in Algorithm 7.4.2, VASS assigns the execution of the longest section type to the best set of cores (those that display lower WUV values), then the exe-

[6]In our applications, it is selected as 2 iterations.

cution of the second longest section type to the next best set of cores, and so on. In other words, VASS performs a one-to-many dynamic pipeline mapping between the section types (i.e., the stages) and the cores such that the overall execution time is reduced. After the section-to-core assignment, once a $\text{core}_i$ encounters a $\text{section}_j$, VASS checks the condition to decide whether $\text{section}_j$ is assigned for the execution on top of $\text{core}_i$. If $\text{section}_j$ is assigned for $\text{core}_i$, it means that there is a match between the characteristics of $\text{core}_i$ and $\text{section}_j$, therefore the execution will be performed. Otherwise VASS does not allocate the $\text{section}_j$ to the $\text{core}_i$. Thanks to the `nowait` statement, for a `parallel sections` consists of $N_{sec}$ sections, VASS replicates the entire `parallel sections` for $R = N_{core}/N_{sec}$ times to maintain all $N_{core}$ cores active while reducing overall pipeline duration.

**Algorithm 7.4.2.** VASS $(sec_0 : sec_{N_{sec}})$

$sortedSecList \leftarrow SortSectionsWUV(sec_0 : sec_{N_{sec}})$
**while** $sortedSecList \neq EMPTY$
$$\textbf{do} \begin{cases} secID \leftarrow extractTopList(sortedSecList) \\ \{coreIDs\} \leftarrow findBestSetCores(secID) \\ tag[\{coreIDs\}] \leftarrow tag[\{coreIDs\}] \bigcup secID \end{cases}$$
**return** $(tag[core_0 : core_{N_{core}}])$

## 7.5 Experimental Results

### 7.5.1 *Framework Setup*

We demonstrate our approach on an OpenMP-enabled SystemC-based virtual platform [28] modeling the tightly coupled cluster described in Sect. 7.2. The virtual platform supports tasking on top of a runtime [24] optimized for the target platform. Table 7.1 summarizes the main architectural parameters, a typical setup for the considered platform template (see [1]). To emulate variations on the virtual platform, we have integrated variations models at the level of individual instructions using the ILV characterization methodology presented in [10]. Integration of ILV models for every core enables online assessment of presence or absence of errant instructions at the certain amount of dynamic voltage and temperature variations. We re-characterized ILV models of an in-order RISC LEON-3 [29] core for 45-nm, for which an advanced open-source RISC core with back-end details for variation analysis is available. First, we synthesized the VHDL code of LEON-3 with the 45-nm TSMC technology library, general-purpose process. The frontend flow with normal $V_{TH}$ cells has been performed using *Synopsys DesignCompiler*, while *Synopsys IC Compiler* has been used for the back-end where the core is optimized for performance.

**Table 7.1** Architectural parameters for VOMP cluster

| | | | |
|---|---|---|---|
| ARM v6 core | 16 | TCDM banks | 16 |
| I$ size | 16KB per core | TCDM latency | 2 cycles |
| I$ line | 4 words | TCDM size | 256KB |
| Latency hit | 1 cycle | L2 latency | ⩾60 cycles |
| Latency miss | ⩾59 cycles | L2 size | 256MB |

To observe the effects of a full range of dynamic voltage and temperature variations, we analyze the delay variability on the individual instructions, leveraging voltage–temperature scaling features of *Synopsys PrimeTime* for the composite current source approach of modeling cell behavior. Finally, delay variability is annotated to the gate-level simulations for creating ILV models. To utilize ILV models on the virtual platform, each core maps ARM v6 instructions to the corresponding ILV models in an instruction-by-instruction fashion during execution of tasks. Therefore, every core will face the errant instructions during work-units execution based on the available amount of variations on the variability-affected cluster. From the same flow we also extract energy models for our cluster architecture.

For the following experiments we consider the cluster with 16 cores. To observe the effect of static process variation on the clock frequency of individual cores within the cluster, we analyze how critical paths of each core are affected due to die-to-die and within-die process parameters variation, following the methodology presented in [12]. Each core maximum frequency varies significantly due to the process variation. As a result, six cores for 16-core cluster cannot meet the design time target clock frequency. To compensate this core-to-core frequency variation, the $V_{DD}$-hopping technique [18] uses the measured delay variation of each core and then selects one of available three discrete voltage modes: $V_{DD}$-high, $V_{DD}$-medium, $V_{DD}$-low. This technique mitigates the core-to-core frequency variations within the variability-affected cluster: six cores are powered up with $V_{DD}$-high, four cores with $V_{DD}$-medium, and six cores with $V_{DD}$-low. This ensures all cores work with the design time target frequency, but they face different error rate based on the instruction type and the operating condition.

### *7.5.2 VOMP Results for Tasking*

We use nine widely adopted computational kernels mainly from the image processing domain that we parallelize using `task` directives. These kernels include *RGB-to-HSV* and *XYZ-to-RGB* for colormap conversions, *Integral image* and *Sobel* for filter operations, *FAST* for corner detection, *Color Tracking*, *Strassen* matrix multiplication, and *Blowfish* for encryption/decryption. Each kernel has one task type, therefore there is no task dependency during execution. We compare the total execution time and energy consumption of VATS, our variability-aware task scheduler, to the baseline RRS policy. Figure 7.12 shows the execution time for all the kernels for three operating corners with temperature of 0, 40, and 80 °C. VATS aims at reducing the instruction replays by allocating tasks on reliable cores while taking into account the load of every queue. As a result, at an operating temperature of 0 °C, VATS achieves up to 30% better performance than RRS, and 13% on average. This clearly indicates that the entire overhead of the variation-tolerant technique is paid off, including the online task characterization, reading and updating WUV metadata, and cost of execution of Algorithm 7.4.1. As shown, VATS displays a robust behavior across a wide range of temperature variations thanks to the reflection by the *always-on* characterizations. At higher temperature, VATS achieves better average performance gain of 17% (at 40 °C) and 21% (80 °C), since WUV is increased at higher temperature.

Figure 7.13 shows the energy consumption of the kernels for VATS normalized to RRS. VATS achieves on average 21% and up to 38% better energy efficiency than RRS at the temperature of 0 °C. VATS further reaches to an average energy saving of 31% at the operating temperature of 80 °C.

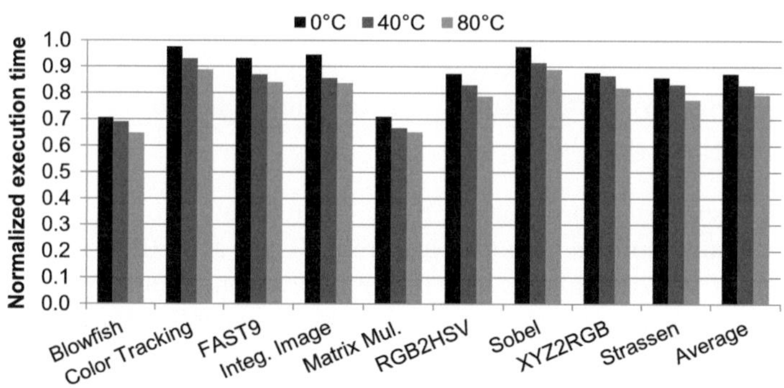


**Fig. 7.12** Execution time for VATS normalized to RRS under temperature variation

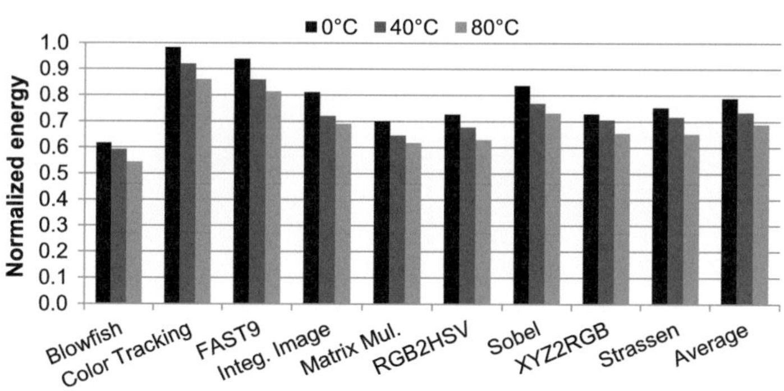


**Fig. 7.13** Energy consumption for VATS normalized to RRS under temperature variation

We also compare the TLV technique with the centralized queue proposed in [9]. TLV, which has variation-agnostic task insertion operations displays on average 75% slower execution than RRS. TLV is on average 100% less energy efficient than RRS. This lack of efficient utilization of resources under variability is mainly because of TLV characterization that does not consider the overall system workload. Its single tasking queue also limits the potentials of task scheduling policies: a core can utilize TLV to only decide whether to proceed to the execution of a task or leave it in the single queue for other cores that leads to an imbalanced system.

### *7.5.3 VOMP Results for Sections*

For evaluating VOMP in the `parallel sections`, we used seven computational intensive kernels amenable to software pipelining. Pitch extractor algorithm (*PEA*), and FFT with covariance matrix factorization (*DFT-COV*) are embedded signal processing kernels extracted from [30, 31]. *Sobel* and *Prewitt* are filter operations useful in the edge detection algorithms. *N-body* is a simulation of a large number of particles under the influence of physical forces. *Mersenne twister* is a pseudorandom number generator. Synthetic is a microkernel implementing a 4-stage parallel pipeline (see Fig. 7.5), representative of streaming applications [32]. We evaluate the effectiveness and robustness of our approach across a wide temperature range of 80 °C.

Figure 7.14 shows the normalized performance (execution time) of VASS to FCFS for three operating corners with temperature of 0, 40, and 80 °C. At an operating temperature of 0 °C, the total execution time is reduced on average by 31% (and up to 40%) thanks to proper assignment of sections to those cores that avoid unbalanced pipelines. This is accomplished by preventing the worst cores from executing a section type that leads to the highest WUV. At the temperature of 80 °C, VASS reaches

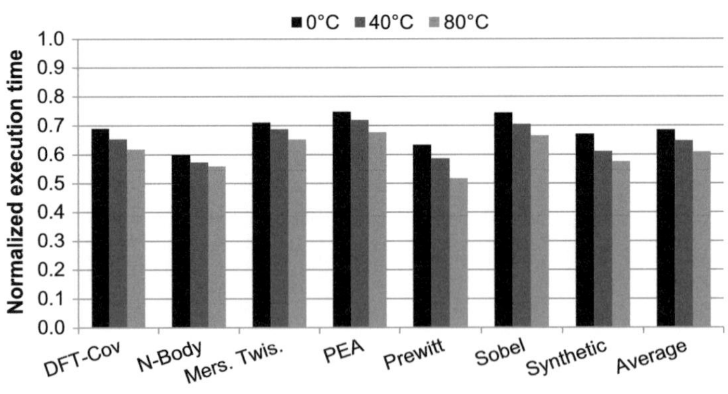


**Fig. 7.14** Execution time for VASS normalized to FCFS under temperature variation

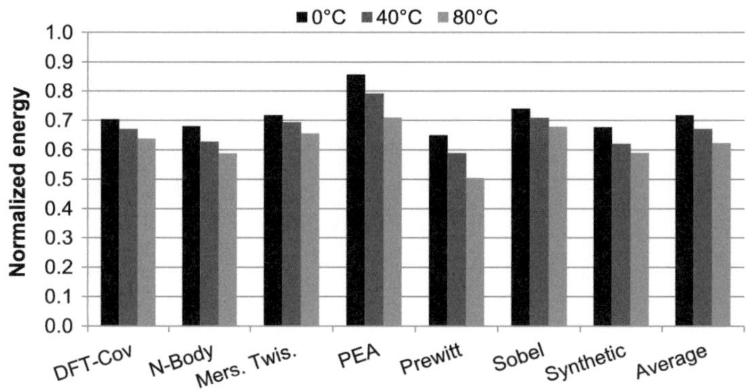


**Fig. 7.15** Energy consumption for VASS normalized to FCFS under temperature variation

on average 39% performance improvement, thanks to the online WUV metadata characterization which reflects the latest temperature variations, thus enabling the scheduler to react accordingly.

Moreover, as shown in Fig. 7.15, VASS simultaneously reduces the total dynamic instruction count that yields an average of 28% (up to 35%) reduction in energy consumption at an operating temperature of 0 °C. A similar pattern for energy saving is observed under temperature fluctuations, confirming the robustness of our approach. VASS reduces energy consumption on average by 37% for high operating temperatures of 80 °C.

## 7.6 Chapter Summary

Circuit failures due to timing errors are considered an important concern in the design of reliable circuits. In this chapter, we show that processing cores can be made robust against an important class of such errors, caused by manufacturing and environmental variabilities, by raising the visibility of such failures across the hardware/software boundary. This is achieved by attaching *metadata* that captures work-unit vulnerability (WUV) from hardware sensing circuits to the runtime system via the software stack. We specifically address its implementation in a parallel execution environment that associates WUV metadata to OpenMP parallel constructs: `task`, `sections`, and `for`. WUV metadata is characterized during work-unit execution on individual cores, and is used to efficiently schedule new instances of the same work-unit type. We have implemented our approach in VOMP, a variability-aware OpenMP execution environment. With VOMP, we propose scheduling algorithms for `tasks` and `sections` that use WUV metadata for countermeasures against variability-induced timing errors. This matches the characteristics of different variability-affected cores to the error vulnerability of different work-unit types in the program, minimizing the need for timing error recovery and the associated costs. Across a wide operating temperature of 80 °C, VOMP effectively eliminates the timing error recovery in the 16-core cluster resulting in average 17 and 36% faster execution for `tasks` and `sections`, respectively. VOMP achieves an average energy saving of 27% for `tasks` and 33% for `sections`.

## References


1. D. Melpignano, L. Benini, E. Flamand, B. Jego, T. Lepley, G. Haugou, F. Clermidy, D. Dutoit, Platform 2012, a many-core computing accelerator for embedded socs: performance evaluation of visual analytics applications, in *Design Automation Conference (DAC), 2012 49th ACM/EDAC/IEEE* (2012), pp. 1137–1142
2. L.M. de Lima Silva, A. Calimera, A. Macii, E. Macii, M. Poncino, Power efficient variability compensation through clustered tunable power-gating. IEEE J. Emerg. Sel. Top. Circuits Syst. **1**(3), 242–253 (2011)
3. K.A. Bowman, J.W. Tschanz, S.L. Lu, P.A. Aseron, M.M. Khellah, A. Raychowdhury, B.M. Geuskens, C. Tokunaga, C.B. Wilkerson, T. Karnik, V.K. De, A 45 nm resilient microprocessor core for dynamic variation tolerance. IEEE J. Solid-State Circuits **46**(1), 194–208 (2011)
4. K.A. Bowman, J.W. Tschanz, N.S. Kim, J.C. Lee, C.B. Wilkerson, S.L. Lu, T. Karnik, V.K. De, Energy-efficient and metastability-immune resilient circuits for dynamic variation tolerance. IEEE J. Solid-State Circuits **44**(1), 49–63 (2009)
5. H. Zakaria, L. Fesquet, Designing a process variability robust energy-efficient control for complex socs. IEEE J. Emerg. Sel. Topics Circuits Syst. **1**(2), 160–172 (2011)
6. P. Gupta, Y. Agarwal, L. Dolecek, N. Dutt, R.K. Gupta, R. Kumar, S. Mitra, A. Nicolau, T.S. Rosing, M.B. Srivastava, S. Swanson, D. Sylvester, Underdesigned and opportunistic computing in presence of hardware variability. IEEE Trans. Comput.-Aided Des. Integr. Circuits Syst. **32**(1), 8–23 (2013)
7. G. Karakonstantis, A. Chatterjee, K. Roy, Containing the nanometer "pandora-box": cross-layer design techniques for variation aware low power systems. IEEE J. Emerg. Sel. Topics Circuits Syst. **1**(1), 19–29 (2011)
8. L. Leem, H. Cho, H.-H. Lee, Y.M. Kim, Y. Li, S. Mitra, Cross-layer error resilience for robust systems, in *2010 IEEE/ACM International Conference on Computer-Aided Design (ICCAD)* (2010), pp. 177–180
9. A. Rahimi, A. Marongiu, P. Burgio, R.K. Gupta, L. Benini, Variation-tolerant openmp tasking on tightly-coupled processor clusters, in *Design, Automation Test in Europe Conference Exhibition (DATE), 2013* (2013), pp. 541–546
10. A. Rahimi, L. Benini, R.K. Gupta, Analysis of instruction-level vulnerability to dynamic voltage and temperature variations, in *Design, Automation Test in Europe Conference Exhibition (DATE), 2012* (2012), pp. 1102–1105
11. A. Rahimi, L. Benini, R.K. Gupta, Application-adaptive guardbanding to mitigate static and dynamic variability. IEEE Trans. Comput. (2013)
12. A. Rahimi, L. Benini, R.K. Gupta, Procedure hopping: a low overhead solution to mitigate variability in shared-l1 processor clusters, in *Proceedings of the 2012 ACM/IEEE International Symposium on Low Power Electronics and Design*, ISLPED '12, ACM, New York, NY, USA (2012), pp. 415–420
13. L. Benini, E. Flamand, D. Fuin, D. Melpignano, P2012: building an ecosystem for a scalable, modular and high-efficiency embedded computing accelerator, in *Design, Automation Test in Europe Conference Exhibition (DATE), 2012* (2012), pp. 983–987
14. Whitepaper. Nvidia's next generation cuda compute architecture: Fermi (2009)
15. Plurality, the hypercore processor. http://www.plurality.com/hypercore.html
16. Kalray, mppa. http://www.kalray.eu/products/mppa-manycore-a-multicore-processors-family-13/

17. A. Rahimi, I. Loi, M.R. Kakoee, L. Benini, A fully-synthesizable single-cycle interconnection network for shared-l1 processor clusters, in *Design, Automation Test in Europe Conference Exhibition (DATE), 2011* (2011), pp. 1–6
18. S. Miermont, P. Vivet, M. Renaudin, A power supply selector for energy- and area-efficient local dynamic voltage scaling, in *Proceedings of the 17th International Workshop on Integrated Circuit and System Design. Power and Timing Modeling, Optimization and Simulation*, PATMOS '07 (Springer, Berlin, 2007), pp. 556–565
19. The gnu project, gomp – an openmp implementation for gcc. http://gcc.gnu.org/projects/gomp
20. E. Ayguade, N. Copty, A. Duran, J. Hoeflinger, Y. Lin, F. Massaioli, X. Teruel, P. Unnikrishnan, G. Zhang, The design of openmp tasks. IEEE Trans. Parallel Distrib. Syst. **20**(3), 404–418 (2009)
21. P.N. Sanda, J.W. Kellington, P. Kudva, R. Kalla, R.B. McBeth, J. Ackaret, R. Lockwood, J. Schumann, C.R. Jones, Soft-error resilience of the ibm power6 processor. IBM J. Res. Dev. **52**(3), 275–284 (2008)
22. G. Hoang, R.B. Findler, R. Joseph, Exploring circuit timing-aware language and compilation, in *Proceedings of the Sixteenth International Conference on Architectural Support for Programming Languages and Operating Systems*, ASPLOS XVI, ACM, New York, NY, USA (2011), pp. 345–356
23. E.K. Ardestani, E. Ebrahimi, G. Southern, J. Renau, Thermal-aware sampling in architectural simulation, in *Proceedings of the 2012 ACM/IEEE International Symposium on Low Power Electronics and Design*, ISLPED '12, ACM, New York, NY, USA (2012), pp. 33–38
24. A. Marongiu, P. Burgio, L. Benini, Fast and lightweight support for nested parallelism on cluster-based embedded many-cores, in *Design, Automation Test in Europe Conference Exhibition (DATE), 2012* (2012), pp. 105–110
25. P. Burgio, G. Tagliavini, A. Marongiu, L. Benini, Enabling fine-grained openmp tasking on tightly-coupled shared memory clusters, in *Design, Automation Test in Europe Conference Exhibition (DATE), 2013* (2013), pp. 1504–1509
26. S.N. Agathos, V.V. Dimakopoulos, A. Mourelis, A. Papadogiannakis, Deploying openmp on an embedded multicore accelerator, in *2013 International Conference on Embedded Computer Systems: Architectures, Modeling, and Simulation (SAMOS XIII)* (2013), pp. 180–187
27. O. Tahan, M. Shawky, Using dynamic task level redundancy for openmp fault tolerance, in *Proceedings of the 25th International Conference on Architecture of Computing Systems*, ARCS'12 (Springer, Berlin, 2012) pp. 25–36
28. D. Bortolotti, C. Pinto, A. Marongiu, M. Ruggiero, L. Benini, Virtualsoc: A full-system simulation environment for massively parallel heterogeneous system-on-chip, in *IPDPS Workshops* (2013), pp. 2182–2187
29. Leon3. http://www.gaisler.com/cms/
30. P.D. Hoang, J.M. Rabaey, Scheduling of dsp programs onto multiprocessors for maximum throughput. IEEE Trans. Signal Process. **41**(6), 2225–2235 (1993)
31. V.K. Prasanna M. Lee, W. Liu, A mapping methodology for designing software task pipelines for embedded signal processing, in *Parallel and Distributed Processing* (1998), pp. 937–944
32. A. Moreno, E. Cesar, A. Guevara, J. Sorribes, T. Margalef, Load balancing in homogeneous pipeline based applications. Parallel Comput. **38**(3), 125–139 (2012)

# Chapter 8
# Memristive-Based Associative Memory for Error Recovery

**Abstract** Thousands of deep and wide pipelines working concurrently make GP-GPU high power consuming parts. Energy-efficient techniques employ voltage overscaling that increases timing sensitivity to variations and hence aggravating the energy use issues. This chapter proposes a method to increase spatiotemporal reuse of computational effort by a combination of compilation and microarchitectural design to enhance error recovery. An associative memristive memory (AMM) module is integrated with the floating point units (FPUs) for *exact* computing. Together, we enable fine-grained partitioning of values and find high-frequency sets of values for the FPUs by searching the space of possible inputs, with the help of application-specific profile feedback. For every kernel execution, the compiler pre-stores these high-frequent sets of values in AMM modules—representing partial functionality of the associated FPU—that are concurrently evaluated over two clock cycles. Our simulation results show high hit rates with 32-entry AMM modules that enable 36% reduction in average energy use by the kernel codes. Compared to voltage overscaling, this technique enhances robustness against timing errors with 39% average energy saving. This proposed method not only reduces energy consumption of error-free operations but also enhances the scope of *error recovery* in a GP-GPU architecture. It is accomplished through an ultra-low-power error recovery via computational reuse, thus offering both scalability and low-cost self-resiliency in the face of high timing error rates. Further, our method leverages memristor technology in the right angle by limiting the stress of write to finite number of write operations only at the start of kernel execution, therefore extending the lifetime of AMM modules. This chapter enhances methods for detecting and correcting the timing errors in GP-GPUs using memristor technology.

## 8.1 Introduction

The scaling of physical dimensions in semiconductor circuits opens the way to an astonishing over 7 billion transistors on a 28 nm process which gives a grand total of 2,880 CUDA cores in recent GP-GPU chips enforcing energy efficiency as a primary concern [1]. Near-threshold computing (NTC) and supply voltage overscaling

 
A. Rahimi et al., *From Variability Tolerance to Approximate Computing in Parallel Integrated Architectures and Accelerators*,
DOI 10.1007/978-3-319-53768-9_8

(VOS) are primary approaches to build energy-efficient circuits [2]. These techniques achieve energy efficiency at a cost to performance. To compensate this performance loss, microarchitectural approach [3] has been proposed to apply these low-power techniques to single instruction multiple data (SIMD) architectures that exploit data parallelism.

Unfortunately, technology scaling also comes with the side effect of ever-increasing parametric variations across process, voltage, and temperature (PVT) [4], which are expected to worsen in future technologies [5]. The most common effect of variability is delay variation that causes circuit-level timing errors. Both NTC and VOS exacerbate the effects of timing errors. Clearly, design methods are needed to make a design resilient to timing errors. Low-voltage resilient technique applies to both logic and memory blocks. For logic, Razor [6] circuit sensors have been employed in the critical paths of the pipeline stages to reduce voltage guardbanding close to *edge-of-failure*. A common strategy is to detect variability-induced delays by sampling and comparing signals near the clock edge to detect the timing errors. The timing errors are then corrected by a recovery mechanism [7]. This recovery process imposes energy overhead and latency penalty of up to 28 extra recovery cycles per error for the seven-stage integer pipeline [7].

In nonvolatile memory area, resistive RAM (ReRAM/memristor) is a promising candidate with fast write speed and low-power operation [8]. To avoid its read disturbance challenge, reliable read operation techniques are proposed including a process-temperature-aware dynamic bit-line bias scheme on a 4-Mb memristor fabricated chip [8]. Li et al. demonstrate a 1-Mb ternary content addressable memory (TCAM) test chip using 2-transistor/2-resistive-phase-change-storage (2T-2R) cells [9]. It achieves $>10\times$ smaller cell size than SRAM-based TCAMs, and ensures reliable low-voltage search operation in the presence of PVT variations, thanks to a clocked self-referenced sensing scheme [9].

For our GP-GPU targets, floating point (FP) pipelines consume higher energy-per-instruction than their integer counterparts and typically have high latency for instance over 100 cycles to execute on a GP-GPU [10]. As energy becomes the dominant design metric, aggressive VOS and NTC increase the rate of timing errors and correspondingly the costs (in energy, performance) of the recovery mechanisms [2, 3]. This cost is exacerbated in FP SIMD architectures where there are wide parallel lanes with deep pipelined stages. This makes the cost of recovery per single error quadratically more expensive relative to scalar functional units [11]. Effectively, the energy-hungry high-latency FP pipelines are prone to inefficiencies under the timing errors.

Parallel execution in the GP-GPU architectures provides an important ability to reuse computation for reducing energy. This chapter exploits this opportunity to make three main contributions:

1. We propose compiler analysis and memristive-based associative memory design (AMM) to identify frequent redundant computations, carefully pre-store these key computations in appropriate AMM, and reuse them to avoid re-executions.

2. To enable spatiotemporal hardware reconfigurability, we tightly integrate an AMM to every FPU in GP-GPUs. The AMM is a software programmable module composed of a TCAM and a crossbar-based memristive memory block that together represent the pre-stored computations as partial functionality of the associated FPU. The AMM module here performs an exact matching during comparisons, and hence does not produce any *intentional* error into the program and maintains 100% numerical correctness. The framework applies a fine-grained value partitioning, and finds high-frequent sets of values for FPUs by searching the space of possible inputs, with the help of application-specific profile feedback described in Sect. 8.3. For every kernel execution, compiler pre-stores these high-frequency sets of values in AMM modules that are concurrently evaluated over two clock cycles, thus creating a spatiotemporal computing model.
3. We demonstrate the effectiveness and robustness of our technique on the Evergreen GP-GPUs. Our experimental results in Sect. 8.4 show that the AMM modules with 32-entry exhibit high hit rate that avoids redundant re-execution by FPUs, therefore resulting in 36% reduction in average energy. Moreover, given that the AMM modules have ample time margins, upon a hit event the likelihood of error recovery is reduced that further improves the energy efficiency. This enhances robustness in VOS scenario with frequent timing errors.

## 8.2 Energy-Efficient GP-GPUs

We focus on the Evergreen family of AMD GP-GPUs (a.k.a. Radeon HD 5000 series), which targets general-purpose data-intensive applications. The Radeon HD 5870 GP-GPU consists of 20 compute units, a global front-end ultra-thread dispatcher, and a crossbar to connect the memory hierarchy. Each compute unit contains a set of 16 Stream Cores (SCs), i.e., 16 parallel lanes. Within a compute unit, a shared instruction fetch unit provides the same machine instruction for all SCs to execute in a SIMD fashion. Each SC contains five Processing Elements (PEs)—labeled X, Y, Z, W, and T—forming an ALU engine to execute Evergreen machine instructions in a vector-like fashion. Finally, the ALU engine has a pool of pipelined integer and FP units. The block diagram of the architecture is shown in Fig. 8.1.

The device kernel is written in OpenCL which runs on a GP-GPU device. An instance of the OpenCL kernel is called a work-item. Each SC is devoted to the execution of one work-item. In the Radeon HD 5870, a *wavefront* is defined as the total number of 64 work-items virtually executing at the same time on a compute unit. To manage 64 work-items in a wavefront on 16 SCs of the compute unit, a wavefront is split into *subwavefronts* at the execute stage, where each subwavefront contains as many work-items as available SCs. In other words, SCs execute the instructions from the wavefront mapped to the SIMD unit in a 4-slot time-multiplexed manner using the integer units and FPUs. The time multiplexing at the cycle granularity relies on the functional units to be fully pipelined.

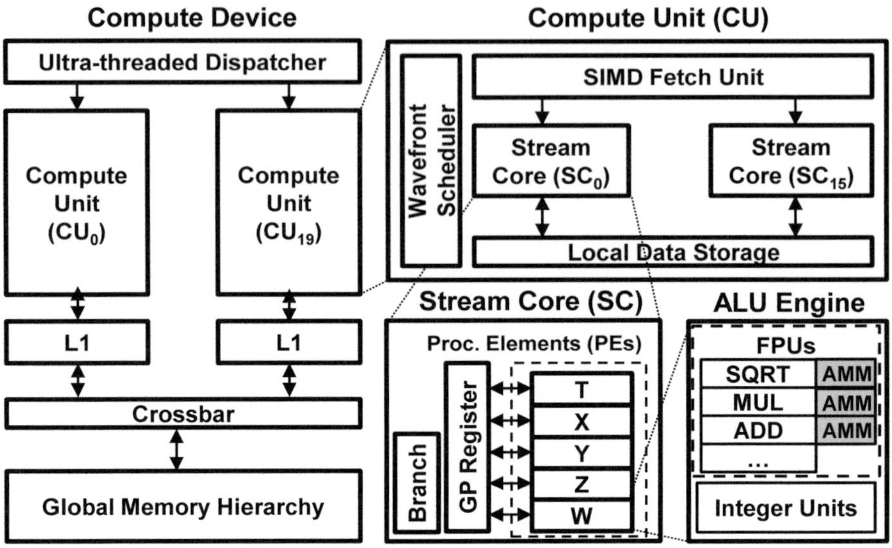


**Fig. 8.1** Block diagram of the Radeon HD 5870 GP-GPU with AMM modules

Evergreen assembly code uses a clause-based format classified into three categories: ALU clause, TEX clause, and control-flow instructions. The control-flow instructions triggering ALU clauses will be placed in the input queue at the ALU engine. There is only one wavefront associated with the ALU engine. After fetch and decode stages, the source operands for each instruction are read that can come from the register file or local memory. For higher throughput, buffers are attached to SCs to read the registers ahead of time. The core stage of a GP-GPU is the execute stage, where arithmetic instructions are carried out in each SC. When the source operands for all work-items in the wavefront are ready, the execution stage starts to issue the operations into the SCs. Finally, the result of the computation is written back to the destination operands.

### *8.2.1 Associative Memristive-Based Computing*

In this subsection we present microarchitectural design of an associative memory module, using memristive parts, that enables partial memory-based computing by leveraging pre-stored high-frequency computations. For every type of FPU, we accordingly designed an AMM module that is tightly integrated to the FPU providing fast local data communication. The key idea is to pre-calculate the output results of a FPU for a partial set of input values and store them before execution on the corresponding AMM module connected to the FPU. In this way, during execution when there is a match between the input values of the FPU and the pre-calculated values, the AMM module returns the pre-stored results on behalf of the FPU at extremely lower energy cost. Therefore, the FPU avoids re-execution and saves energy. The AMM module has a standard interface as it mimics the partial functionality of the associated FPU: as the inputs, it accepts the input operands of the FPU, and as the output it returns the result as well as a hit signal.

The AMM module is composed of two pipelined stages. In the first stage, a TCAM searches the input operands and determines whether there is a match (i.e., hit) between the input operands and the content of TCAM. In the second stage, a 1T-1R memristive memory is used to return the pre-stored output result in case of a match. For TCAM design, we use a memristive 2T-2R cell structure proposed in [9]. Each line in the TCAM stores one set of the frequent input operands, and each bit cell consists of two memristive elements to store the pattern and two access transistors, as

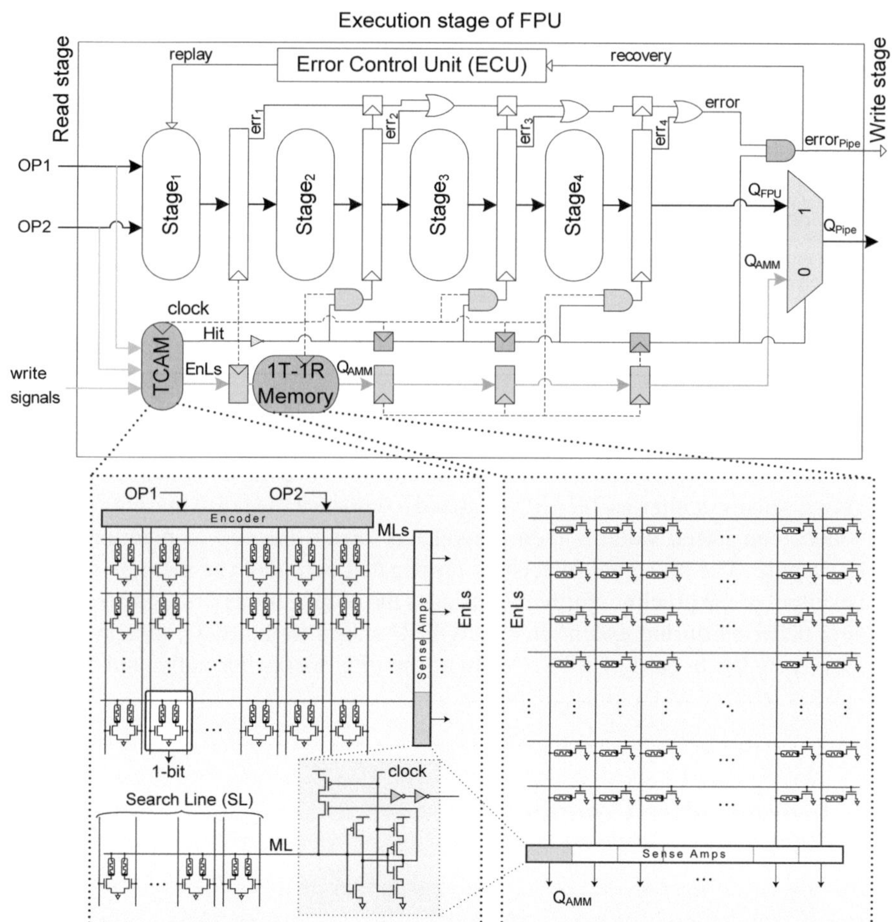


**Fig. 8.2** Execution stage of the FPU with AMM module

shown in Fig. 8.2. To program the TCAM, the write voltages are applied on the match lines (ML), and access transistors of select devices are connected via the search line (SL) to perform the write operation. In order to search the TCAM, match lines are precharged during the precharge phase while all the SLs are inactive to disconnect the access transistors. In the evaluation phase, based on the pattern-under-search, one of two access transistors in each bit cell is ON, connecting the corresponding memristor to the ML. In case of a bit mismatch, ML will be connected to the ground through a low-resistance memristive device. Thus even one bit of mismatch can quickly discharge the ML. In case of a match for a line, the ML is not connected to the ground because of the high-resistant memristive devices and stays at the precharged value for a longer time, providing a clear margin. A clocked self-referenced sensing scheme as well a 2-bit encoding is also applied to further increase the noise margin [9], and provide digital match/mismatch outputs that are fed to the next stage as the enable lines (EnL) which display a one-hot encoding; therefore, the hit signal is the logical OR of EnLs.

In case of a match, the hit signal alongside with the previously computed result ($Q_{AMM}$) are propagated toward the end of the pipeline. TCAM raises the hit signal that squashes the remaining stages of the FPU to avoid the redundant computation by clock-gating; the clock-gating signal is forwarded to the rest of FPU stages, cycle by cycle. Given that the first stage of the FPU is concurrently working with TCAM, considerable energy is saved by spontaneously clock-gating the remaining stages. Instead, the pre-stored result is read from the memristive memory at negligible energy cost. Figure 8.2 shows the structure of such 1T-1R memory that is used to store the output patterns. To program the memory, write voltage is applied on the bit-lines, while the enable lines are used to select the target cell. For read operation, the enable lines are derived by the EnL values of TCAM, thus either none or only one of the enable lines are active at any given clock cycle, connecting a memristive cell to the bit-line. The bit-lines that are precharged during a precharge phase will discharge/remain charged based on the resistance of the connected memristive cell. The same sense circuitry as TCAM is utilized to enhance the noise margins and read the value. The stored value is then propagated toward the end of pipeline for the reuse purpose. The hit signal selects the propagated output of the memory ($Q_{AMM}$) as the output of the pipeline; further, it disables the propagation of timing error signal (if any) occurred during execution of any FPU stages to the ECU, thus avoids the recovery penalty. In case of a TCAM miss, the FPU works normally, and its result ($Q_{FPU}$) is selected as the pipeline output.

## 8.3 Collaborative Compilation

We briefly describe proposed collaborative compiler analysis followed by an evaluation of how memristive-based computing can increase the energy efficiency of GP-GPUs. Figure 8.3 illustrates the collaborative compilation flow. In the profiling stage, we have an OpenCL kernel with a training input dataset. We focus on the indi-

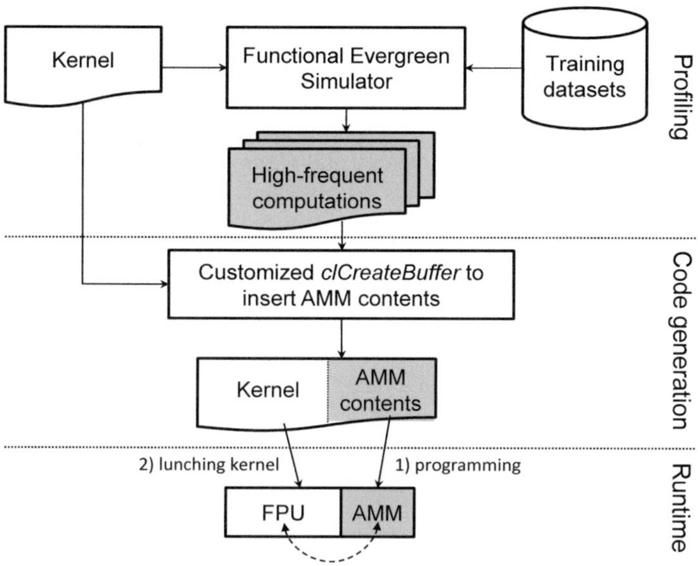


**Fig. 8.3** Collaborative compilation framework and memristive-based computing flow

vidual FPUs to observe the dispersion of the input operands at the finest granularity. To expose high-frequent set of operands for each FP operation, we individually profile every type of FP operation and keep the distinct sets of the input operands and the related result. The kernel is instrumented on the Evergreen functional simulator—this can also be done by proper emission of instrumentation APIs in the naive kernel code. The output of this stage for every FP operation is *high-frequent computations*: a list of top sets of values, i.e., the operands and the related result, which are sorted based on their frequency of occurrence. This profiling stage is a *one-off* activity whose cost is amortized across all future usage of the kernel.

In the next step, the compiler generates codes to store a subset of these high-frequent computations as the content of AMM modules. To do so, the compiler leverages AMD compute abstraction layer (CAL) APIs that facilitate programming AMM modules that are addressable by software. CAL provides a runtime device driver library that supports code generation, kernel loading, and execution, and allows the host program to interact with the stream cores at the lowest level. Right before lunching kernel execution, compiler inserts codes for *programming AMM modules*: for every type of FP operation executed during the kernel, a custom version of "clCreateBuffer" writes the AMM contents (up to few hundred bytes) to the AMM modules accordingly. In this way, we concurrently program all AMM modules integrated to a type of FPU across all PEs in GP-GPUs since their content is equivalent.

### 8.3.1 FPU Memristive-Based Computing

We evaluate the memristive-based computing at the fine-grained instruction-level across all types of the FPUs activated during the execution of two kernels: *Sobel* filter from image processing applications and *Haar* wavelet transform from signal processing applications—more kernels are evaluated in Sect. 8.4.2. Figure 8.4 shows the train and test images for *Sobel* filter. To identify the high-frequent computations, the compiler profiles *Sobel* kernel with the train input image. Four types of FP operations, including addition, multiplication, square root, and multiplication–addition, are activated during the kernel execution; profiler sorts each type and stores top-32 sets with highest frequency of occurrence as AMM contents. Later, for the consecutive kernel executions, the compiler first programs the AMM modules with the

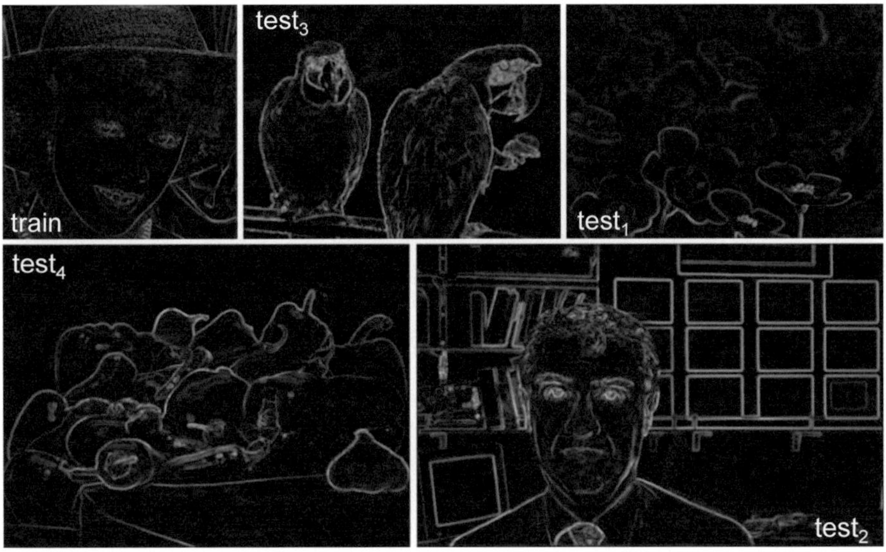


**Fig. 8.4** Train and test images for *Sobel* filter

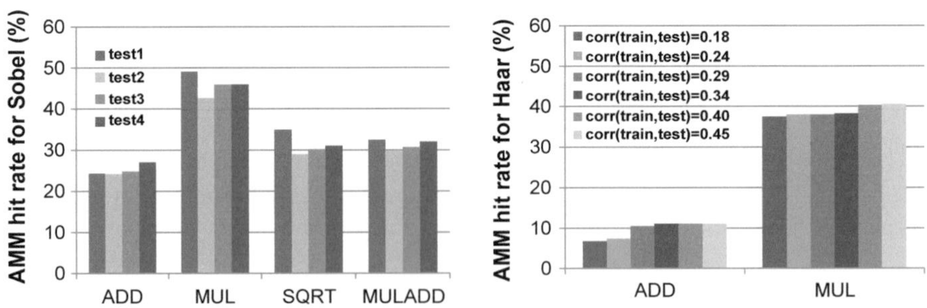


**Fig. 8.5** AMM (32-line) hit rates for (i) *Sobel* with the test images; (ii) *Haar* with various signals

stored AMM contents, and then starts kernel execution. Figure 8.5 shows the AMM hit rates for the activated FP operations during *Sobel* execution with the test images. As shown, the hit rate depends on the FPU operations, but all AMM modules display a hit rate of greater than 25% with a tiny TCAM of 32 lines. The AMM modules for MUL and SQRT exhibit a significant hit rate of up to 49 and 35%, respectively. Overall, an average hit rate of 25, 46, 31, and 31% is observed for ADD, MUL, SQRT, and MULADD, respectively. This means a significant number of operands are matched with the stored computation in the AMM modules, and therefore there is no need for re-executing those values.

To evaluate *Haar* kernel, we use a random signal as the training input and then six different signals having various correlations with the trained input signal. Figure 8.5 shows that the AMM modules display a hit rate in the range of 7–11% for ADD, and 39–41% for MUL. We also evaluate the tradeoff between the hit rate and energy when the AMMs utilizing larger TCAM and memory with 64, 128, and 256 lines. The hit rate of the kernels increases less than 10% when the number of lines is increased from 32 to 256. On the other hand, the AMMs with 32-line display higher energy efficiency (7× higher hit rate per power compared to the AMMs with 256 lines). Therefore, we have used the AMMs with 32-line for our proposed framework, and we also measured its energy efficiency in Sect. 8.4.2. Please note that the AMM content per each kernel occupies few kilobytes, for instance $32 \times 48 = 1.5$ KB for *Sobel*, and $32 \times 24 = 0.75$ KB for *Haar*.

## 8.4 Experimental Results

Our methodology uses the AMD Evergreen GP-GPUs, but can be applied to other GP-GPUs as well. We have selected applications from AMD APP SDK v2.5 [12] in OpenCL. We have examined three image processing filters: *Sobel*, *Gaussian*, and *URNG*; as well as one-dimensional *Haar* wavelet transform, *FastWalsh* transform, *Prefixsum*, and *Eigenvalues* of a symmetric matrix. Multi2Sim [13], a cycle-accurate CPU–GPU simulation framework, is used for profiling. The naive binaries of the kernels are run on the simulator; the input values for the kernels are generated by the default OpenCL host program. We analyzed the effectiveness of the proposed technique in the presence of timing errors and VOS in TSMC 45 nm.

### *8.4.1 FPUs with AMM Modules*

Since the fetch and decode stages display a low criticality [14], we focus on the execution stage consisting of six frequently exercised FPUs: ADD, MUL, SQRT, RECIP, MULADD, and FP2FIX. On Evergreen, every ALU functional unit has a latency of four cycles and a throughput of one instruction per cycle [15]. Therefore, VHDL codes of the FPUs are generated and optimized using FloPoCo [16]—an

**Table 8.1** Energy(pJ) comparison of the FPUs with corresponding AMMs

| | ADD | MUL | SQRT | RECIP | MADD | F2FIX |
|---|---|---|---|---|---|---|
| FPU | 5.81 | 12.76 | 16.92 | 30 | 21.21 | 3.04 |
| AMM | 1.66 | 1.66 | 1.30 | 1.30 | 1.99 | 1.30 |

arithmetic synthesizable FP core generator. To achieve a balanced clock frequency across the FP pipelines, the RECIP has a latency of 16 cycles, while the rest of the FPU have four cycles latency.

The FPUs are synthesized and mapped using the TSMC 45-nm technology library. The front-end flow has been performed using *Synopsys Design Compiler* with the topographical features, while *Synopsys IC Compiler* has been used for the back-end. The design has been optimized for a signoff clock period of 2ns at (SS/0.81 V/125 °C), and then optimized for power. The AMM module has different sizes based on the type of FPU, and its TCAM has 32 × 32 for SQRT, RECIP, and FP2FIX; 32 × 64 for ADD, and MUL; 32 × 96 for MULADD. The transistor-level CMOS circuitry is implemented and then SPICE simulations are done using *Cadence Virtuoso*. For line resistances and capacitances, the same model and numbers used in [17] were assumed. The memristor models are having 250 K Ron and 100 M Roff, and are based on the fabricated memristors in [18]. To integrate the resilient architecture, the AMM modules are integrated into the FPUs pipelines with the multiple-issue recovery mechanism [7].

Table 8.1 summarizes the power results of FPUs and AMMs' implementations. As shown, integration of FPUs with AMMs incurs negligible overhead and it is entirely paid off by the power saving due to the frequent clock-gating of the FPUs during the hit events that results into even higher energy efficiency detailed in the following subsection. We note that the overhead will be further reduced for deeper pipelines. The AMM module does not limit the clock frequency as it has a positive slack of 300 ps.

### 8.4.2 Energy Saving

We measure the overall AMM modules' hit rates for the image processing filters using two datasets: $dataset_1$ which is a relatively small dataset of ~400 face images [19]; $dataset_2$ which a large 2,000 Web faces [20]. For profiling, we have used only 20 random images from $dataset_1$ as the training inputs. Figure 8.6 shows the worst, the best, and average hit rates for the two datasets. The best hit rate of 84% is observed during *Sobel* execution for one of the images in $dataset_2$. As shown, for every filter, the average hit rate is almost equal across the two different datasets: 38 or 36% for *URNG*, 22 or 24% for *Gaussian*, and 34% for *Sobel*. The worst hit rate is 13% that *Gaussian* filter experienced in one of the images in the large $dataset_2$, guaranteeing the

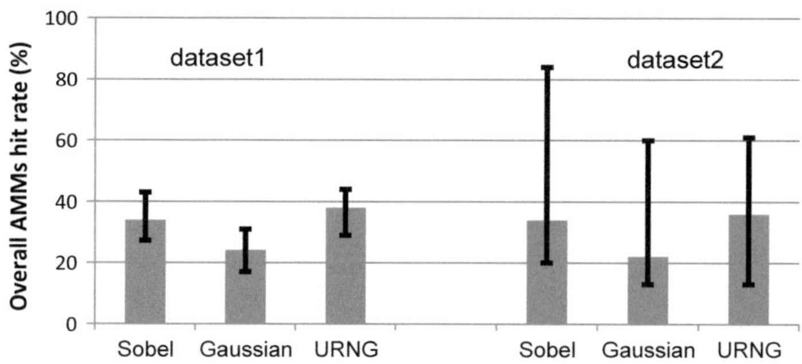


**Fig. 8.6** Overall AMM hit rates for test datasets: $dataset_1$ [19], $dataset_2$ [20]

absence of a poor locality in real-life datasets. It therefore confirms the applicability of profiling for the associative memory-based computing. The proposed optimization framework is based on either profiling or designer knowledge (provided from a domain expert). We should note that the profiling is a common technique used for runtime optimizations [21].

We evaluate the energy saving of our proposed architecture with a baseline architecture that utilizes recent resilient techniques: Razor error detection [6], and the scalable recovery mechanism of the multiple-issue instruction replay [7] adapted for the FPUs. Our architecture (FPUs + AMMs) superposes the AMM modules on the baseline architecture. Figure 8.7 illustrates the energy consumption of the two architectures at different voltage points for each kernel. At the nominal voltage of 1.0 V, where there is no timing errors, the proposed architecture with AMM modules achieves 36% better energy efficiency across all the kernels, thanks to the high hit rates in the AMMs. This is accomplished through the appropriate coupling of the memristive-based computing and value prediction that is extended to GP-GPU architectures.

We also assess the efficacy of the proposed architecture in the VOS regime while clocking at constant speed. To do so, the voltage of FPUs is scaled down in the range of 1.0–0.88 V. To ensure always correct functionality of the AMM modules, we maintain their operating voltage at the fixed nominal 1.0 V. We employ voltage scaling feature of *Synopsys PrimeTime* to analyze the delay variations under the voltage overscaling. Then, the voltage overscaling-induced delay is back annotated to the post-layout simulation which is coupled with Multi2Sim simulator to quantify the timing error rate. The baseline architecture triggers the recovery mechanism when any voltage overscaling-induced timing error occurs, while our proposed architecture does it in case of simultaneous events of the error and the AMM miss.

At the nominal voltage of 1.0 V, without any timing error, the proposed architecture reaches up to 76% energy saving for *FastWalsh*. The proposed architecture also exhibits a great potential of survival in the VOS regime. Scaling down the voltage below 0.92 V for the FPUs causes abrupt increasing of the error rate and therefore these units incur frequent recovery cycles. Our implementation excludes the fact

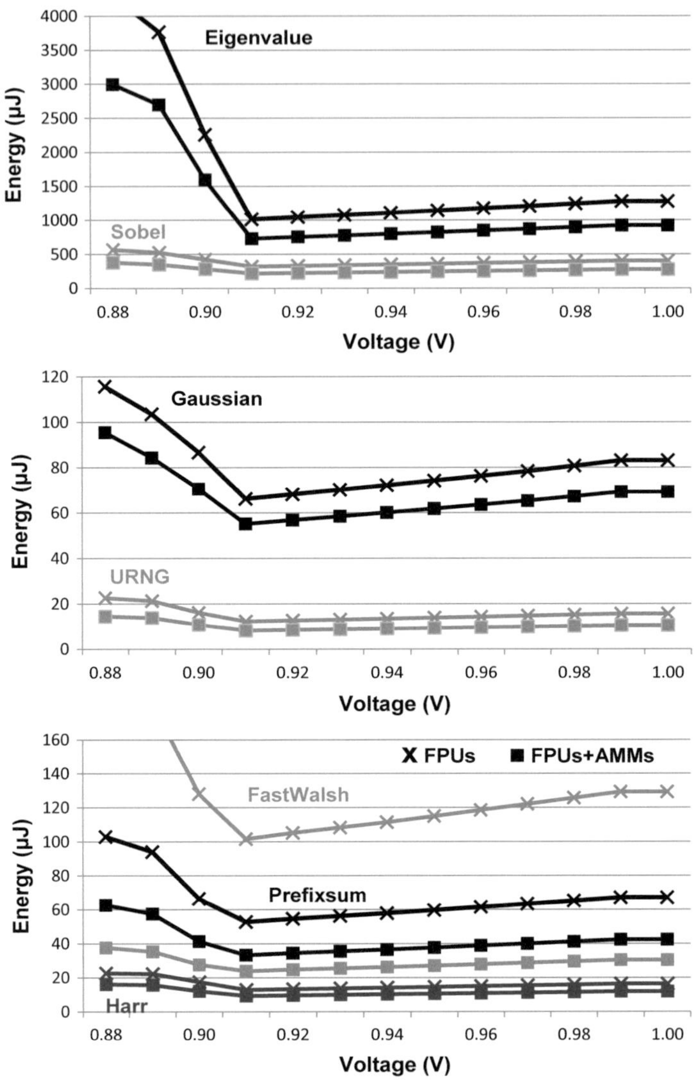


**Fig. 8.7** Total energy consumption of proposed architecture with AMM modules (FPUs + AMMs) in comparison with the baseline architecture (FPUs) under VOS

that the AMM module may produce an erroneous result, because the module has a positive slack of 300 ps and always works at the nominal voltage proving sufficient guardband. Therefore, it is unlikely for AMM modules to face any *timing* errors. In the voltage range of 0.92–0.88 V, the kernels face 10–38% error rate in the baseline architecture which is further reduced to a range of 3–24% in the proposed architecture. The proposed architecture consumes a little bit more energy till 0.88 V because of the errors that are not masked by our AMM modules; it reaches an average energy saving of 39% at voltage of 0.88 V. This is accomplished through the efficient timing error recovery by associative memristive-based modules that do not impose any penalty as opposed to the baseline recovery.

## 8.5 Chapter Summary

This chapter proposes static compiler analysis and coordinated microarchitectural design that enable efficient reuse of computations in GP-GPUs. The proposed technique makes use of emerging associative memristive modules connected with floating point units that enables spatial and temporal computational reuse. Fast and efficient accesses to the pre-stored computation are guaranteed by carefully placing these key values in tightly coupled associative memory modules. The GP-GPU kernels exhibit a low entropy that is high contextual information, yielding up to 84% hit rate on the 32-entry AMMs with an average energy saving of 36%. Our proposed framework also enhances robustness and energy saving in the VOS regime by avoiding conventional timing error recovery costs. This technique highly surpasses the baseline architecture by an average energy saving of 39%.

## References


1. Whitepaper. NVIDIAs next generation CUDATM compute architecture: Kepler TM GK110 (2012)
2. D. Jeon, M. Seok, Z. Zhang, D. Blaauw, D. Sylvester, Design methodology for voltage-overscaled ultra-low-power systems. IEEE Trans. Circuits Syst. II Express Briefs, **59**(12), 952–956 (2012)
3. R. Pawlowski, E. Krimer, J. Crop, J. Postman, N. Moezzi-Madani, M. Erez, P. Chiang, A 530mv 10-lane SIMD processor with variation resiliency in 45nm SOI, in *2012 IEEE International Solid-State Circuits Conference Digest of Technical Papers (ISSCC)* (2012), pp. 492–494
4. S. Borkar, T. Karnik, S. Narendra, J. Tschanz, A. Keshavarzi, V. De, Parameter variations and impact on circuits and microarchitecture, in *Proceedings of Design Automation Conference, 2003* (2003), pp. 338–342
5. The ITRS website. http://www.itrs.net/Links/2011ITRS/Home2011.htm
6. S. Das, D. Roberts, S. Lee, S. Pant, D. Blaauw, T. Austin, K. Flautner, T. Mudge, A self-tuning DVS processor using delay-error detection and correction. IEEE J. Solid-State Circuits, **41**(4), 792–804 (2006)
7. K.A. Bowman, J.W. Tschanz, S.L. Lu, P.A. Aseron, M.M. Khellah, A. Raychowdhury, B.M. Geuskens, C. Tokunaga, C.B. Wilkerson, T. Karnik, V.K. De, A 45 nm resilient microprocessor core for dynamic variation tolerance. IEEE J. Solid-State Circuits **46**(1), 194–208 (2011)
8. M.-F. Chang, S.-S. Sheu, K.-F. Lin, C.-W. Wu, C.-C. Kuo, P.-F. Chiu, Y.-S. Yang, Y.-S. Chen, H.-Y. Lee, C.-H. Lien, F.T. Chen, K.-L. Su, T.-K. Ku, M.-J. Kao, M.-J. Tsai, A high-speed 7.2-ns read-write random access 4-mb embedded resistive RAM (ReRAM) macro using process-variation-tolerant current-mode read schemes. IEEE J. Solid-State Circuits **48**(3), 878–891 (2013)

9. J. Li, R.K. Montoye, M. Ishii, L.Chang, 1 Mb 0.41$\mu$m$^2$ 2T-2R cell nonvolatile TCAM with two-bit encoding and clocked self-referenced sensing. IEEE J. Solid-State Circuits **49**(4), 896–907 (2014)
10. Micro-benchmarking the GT200 GPU. Technical report, Computer Group, ECE, University of Toronto
11. A. Rahimi, L. Benini, R.K. Gupta, Temporal memoization for energy-efficient timing error recovery in GPGPUs, in *Design, Automation and Test in Europe Conference and Exhibition (DATE), 2014* (2014), pp. 1–6
12. AMD APP SDK v2.5. http://www.amd.com/stream
13. Multi2Sim: a heterogeneous system simulator. https://www.multi2sim.org/
14. A. Rahimi, L. Benini, R.K. Gupta, Analysis of instruction-level vulnerability to dynamic voltage and temperature variations, in *Design, Automation Test in Europe Conference Exhibition (DATE), 2012* (2012), pp. 1102–1105
15. AMD evergreen family instruction set architecture (2011)
16. Flopoco: floating-point cores generator. http://flopoco.gforge.inria.fr/
17. A. Ghofrani, M.A. Lastras-Montano, K.-T. Cheng, Towards data reliable crossbar-based memristive memories, in *2013 IEEE International Test Conference (ITC)* (2013), pp. 1–10
18. K.-H. Kim, S. Gaba, D. Wheeler, J.M. Cruz-Albrecht, T. Hussain, N. Srinivasa, L. Wei, A functional hybrid memristor crossbar-array/CMOS system for data storage and neuromorphic applications. Nano Lett. **12**(1), 389–395 (2012) (PMID: 22141918)
19. Caltech 101 dataset. http://www.vision.caltech.edu/Image_Datasets/Caltech101/
20. Caltech 10K web faces dataset. http://www.vision.caltech.edu/Image_Datasets/Caltech_10K_WebFaces/
21. H. Esmaeilzadeh, A. Sampson, L. Ceze, D. Burger, Neural acceleration for general-purpose approximate programs, in *2012 45th Annual IEEE/ACM International Symposium on Microarchitecture (MICRO)* (2012), pp. 449–460

# Part III
# Accepting Errors

In this part, we describe new methods that have a relaxed behaviour toward the timing error handling. The methods presented in the previous parts strive to achieve instruction executions exactly as specified by the application programs. In contrast, probabilistic or approximate programs can exhibit enhanced error resilience for applications when multiple valid output values are permitted. Conceptually, such programs have 'elastic outputs', and if execution is not 100% numerically correct, the program can still appear to execute correctly from the user's perspective. Programs with elastic outputs have application-dependent fidelity metrics, such as peak signal to noise ratio, associated with them to measure the quality of the computational result. The degradation of output quality for such applications is acceptable if the fidelity metrics satisfy a certain threshold. This provides an opportunity for ignoring the effect of timing errors as long as such errors do not lead to program failures, crashes, or hangs. Table 5 illustrates these techniques with spacial emphasis on application/algorithm, architecture, and circuit levels.

This part explores the possibility and consequences of accepting errors, or what is called "approximate computing" paradigm [138, 134]. In other words, we seek ways for continued operation of a computer system even in the presence of errors. In Chap. 9, we propose programming and runtime environment to support controlled approximate computing in tightly-coupled shared-memory processor clusters. It provides OpenMP extensions as custom directives for floating-point computations to specify parts of a program that can be executed "approximately". Using the notions of approximate and exact computing, we have built a compiler and architecture environment to use approximate computations in a user- or algorithmically-controlled fashion. This is achieved via design-time profiling, synthesis, and optimization in conjunction with runtime characterization techniques. This approach eliminates the cost of error correction for specific annotated regions of code if and only if the propagated error significance and error rate meet application-specific constraints on quality of output. At design-time, these code regions are profiled to identify acceptable error significance and error rate. This application-specific information drives optimizations for approximate hardware synthesis of floating-point units. At runtime, as different sequences of OpenMP directives are dynamically encountered during program execution, the scheduler promotes the floating-point unit to exact mode, or demotes them to approximate mode depending upon the code region requirements. In addition, in Chap. 10, we also explore purely software transformation methods to unleash untapped capabilities of the contemporary fabrics for exploiting approximate computing. Exploiting this opportunity is particularly important for field-programmable gate array (FPGA) accelerators that are inherently subject to many resource constraints. To better utilize the FPGA resources, we develop an automated design workflow for FPGA accelerators that leverages approximate computation to increase data-level parallelism and achieve higher computational throughput.

As a follow up to our earlier use of memristive associative memory modules to reduce the cost of error recovery and speed up computations, we also seek its use in approximate computing in Chap. 11. These memristive modules provide an average energy saving of 32% by operating at low-voltages and approximately recalling the frequent computations, hence avoiding re-executions. The modules accept a Hamming distance range of 0-2 during approximate matches that leads to a controllable approximate computing suitable for GPU applications.

GPGPUs execute workload in SIMD fashion with high utilization. We show in Chap. 12, parallel execution in such SIMD architectures provides an important ability to reuse computation (i.e., spatial and temporal memoization).

We conclude with an outlook for the emerging field in Chap. 13.

# Chapter 9
# Accuracy-Configurable OpenMP

**Abstract** We propose an OpenMP programming environment for fine-grained approximate computing on multi-core cluster architecture with shared and accuracy-reconfigurable floating-point units (FPUs). This shared-FPUs cluster dynamically characterize FP pipeline vulnerability (FPV) and expose it as metadata to a software scheduler for reducing the cost of error correction. To further reduce this cost, our programming, and runtime environment also supports *controlled approximate computation* through a combination of design-time and runtime techniques. We provide OpenMP extensions (as custom directives) for FP computations to specify parts of a program that can be executed approximately. We use a profiling technique to identify tolerable error significance and error rate thresholds in error-tolerant image processing applications. This information further guides an application-driven hardware FPU synthesis and optimization design flow to generate efficient FPUs. At runtime, the scheduler utilizes FPV metadata and promotes FPUs to accurate mode, or demotes them to approximate mode depending upon the code region requirements. We demonstrate the effectiveness of our approach (in terms of energy savings) on a 16-core tightly coupled cluster with eight shared-FPUs for both error-tolerant and general-purpose error-intolerant applications. This chapter provides a method for accepting errors in tightly coupled processor clusters with shared FPUs.

## 9.1 Introduction

The cost of error recovery mechanisms is high in the face of frequent timing errors in aggressive voltage down-scaling and near-threshold computation in an attempt to save power [1, 2]. This cost is exacerbated in floating-point (FP) pipelined architectures because FP pipelines typically have high latency, e.g., up to 32 cycles to execute depending upon the type and precision on an ARM Cortex-A9, and higher energy-per-instruction costs than their integer counterparts. Further, deeper pipelines induce higher pipeline latency and higher cost of recovery through flushing and replaying. These energy-hungry high-latency pipelines are prone to inefficiencies under timing errors because the number of recovery cycles per error is increased at least linearly with the pipeline length. More importantly, FP pipelines are often shared among

 
A. Rahimi et al., *From Variability Tolerance to Approximate Computing in Parallel Integrated Architectures and Accelerators*,
DOI 10.1007/978-3-319-53768-9_9

cores due to their large area and power cost. For instance, the AMD Bulldozer architecture shares a floating-point unit (FPU) between a dual-clustered integer core, with four pipelines. UltraSPARC T1 also has a shared-FPU between eight cores. This makes the cost of recovery even more pronounced for a cluster of tightly coupled processors utilizing shared resources.

We present techniques to enhance OpenMP and the shared-memory architecture to support approximate computing. Our goal is to reduce the cost of a resilient FP environment which is dominated by the error correction. Tolerance to error in execution is often a property of the application: some applications, or their parts, are tolerant to errors (notably, media processing applications), while some other parts must be executed exactly as specified. We either explicitly accept the timing errors – if possible – in a fully *controlled* manner to avoid undefined behavior of programs; or we try to reduce the frequency of timing errors by assigning computations to appropriate pipelines with lower vulnerability. Accordingly, this chapter makes three contributions:

1. We propose a set of accuracy-reconfigurable FPUs that are resistant to variation-induced timing errors and shared among tightly coupled processors in a cluster. This resilient shared-FPUs architecture supports online timing error detection, correction, and characterization. We introduce the notion of FP pipeline vulnerability (FPV), captured as metadata, to expose variability and its effects to a software scheduler for reducing the cost of error correction. A runtime ranking scheduler utilizes the FPV metadata to identify the most suitable FPUs for the required computation accuracy for the minimum timing error rate.
2. Using the notions of approximate and accurate computations, we describe a compiler and architecture environment to use approximate computations in a user- or algorithmically controlled fashion. This is achieved via design-time profiling, synthesis, and optimization in conjunction with runtime characterization techniques. This approach eliminates the cost of error correction for specific annotated approximate regions of code if and only if the propagated error significance and error rate meet application-specific constraints on quality of output. For error-tolerant applications our OpenMP extensions specify parts of a program that can be executed approximately, thus providing a new degree of scheduling flexibility and error resilience. At design-time, code regions are profiled to identify acceptable error significance and error rate. This information drives synthesis of an application-driven hardware FPU. At runtime, as different sequences of OpenMP directives are dynamically encountered during program execution, the scheduler promotes FPUs to accurate mode, or demotes them to approximate mode depending upon the code region requirements.
3. Our approach enables efficient execution of finely interleaved approximate and accurate operations enforced by various computational accuracy demands within and across applications. We demonstrate the effectiveness of our approach on a 16-core tightly coupled cluster in the presence of timing errors. For general-purpose error-intolerant application, our approach reduces the recovery cycles that yield an average energy saving of 22% (and up to 28%), compared to the

worst-case design. For error-tolerant image processing applications with annotated approximate directives, 36% energy saving is achieved while maintaining acceptable quality degradation.

## 9.2 Controlled Approximation

Approximate computation leverages the inherent tolerance of some (type of) applications within certain error bounds that are acceptable to the end application. Two metrics have been previously proposed to quantify tolerance to errors [3]: error rate and error significance. The error rate is the percentage of cycles in which the computed value of a FP operation is different from the correct value. The error significance is the numerical difference between the correct and the computed results.

Disciplined approximated programming allows programmers to identify parts of a program for approximate computation [4]. This is commonly found in applications in vision, machine learning, data analysis, and computer games. Conceptually, such programs have a vector of 'elastic outputs' than a singular correct answer. Within the range of acceptable outputs, the program can still appear to execute correctly from the user's perspective [4–6] even if the individual computations are not exact. Programs with elastic outputs have application-dependent fidelity metrics, such as peak signal to noise ratio (PSNR), associated with them to characterize the quality of the computational result. The degradation of output quality for such applications is acceptable if the fidelity metrics satisfy a certain threshold. For example, in multimedia applications the quality of the output can be degraded but acceptable within the constraints of PSNR $\geq$30 dB.

The timing error must be controllable because it could occur anytime and anywhere in the circuit. Therefore, three conditions must be satisfied to ensure that it is safe not to correct a timing error when approximating the associated computation:

1. The error significance is controllable and below a given threshold;
2. The error rate is controllable and below a given error rate threshold;
3. There is a region of the program that can produce an acceptable fidelity metric by tolerating the uncorrected, thus propagated, errors with the above-mentioned properties.

These conditions can be satisfied either through a set of profiling phases, or a set of threshold values specified by a domain expert via application knowledge. As we will detail in Sect. 9.4.1.2, the output information of our profiling phase is a set of threshold values that guarantee an acceptable fidelity metric. Any timing error greater than the set of thresholds triggers the recovery mechanism during the approximate operation to avoid unacceptable accuracy and undefined program behavior (e.g., in case of data-dependent control-flow), therefore guaranteeing a safe approximate computation.

In the following sections, we describe how we use these rules in OpenMP environment to ensure that approximate computations always deliver the required accuracy, and how they can be used for efficient hardware FPU synthesis and optimizations.

## 9.3 Accuracy-Configurable OpenMP Environment

### *9.3.1 Accuracy-Configurable FPUs*

We extend the baseline cluster architecture with our resilient shared-FPUs. Similar to the DMA, our FPU design is also controlled via memory-mapped registers, accessible through a slave port on the peripheral interconnect. As shown in the rightmost part of Fig. 9.1, the FPU has three pipeline blocks which work in parallel. Each pipeline's inputs and outputs are retrieved from a minimal register file (one register file per pipeline to allow for parallel execution). For each pipeline there is a write-only *opmode* register that determines whether the current operation is accurate (i.e., exact) or approximate. Every pipeline block has two dynamically reconfigurable operating modes: (i) accurate, and (ii) approximate. To ensure 100% timing correctness in the accurate mode, every pipeline uses the EDS sensors as well as the ECU to detect and correct any timing errors. During the accurate operation if a timing error is detected, the EDS circuits prevent pipeline from writing results to the register and thus avoid corrupting the architectural state. To recover the errant operation, the ECU adopts the multiple-issue operation replay mechanism [7].

In the approximate mode, the pipeline simply disables the EDS sensors on the less significant *N* bits of the fraction. The sign and the exponent bits are always protected by EDS. This allows the pipeline to ignore any timing error below the less significant *N* bits of the fraction and save on the recovery cost. We only disable the error detection circuits partially on *N* bits of the fraction. This enables the FP pipeline

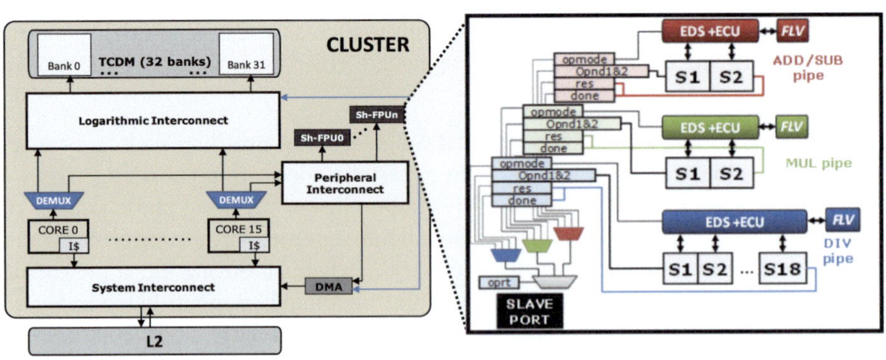


**Fig. 9.1** Variability-aware cluster architecture with accuracy-configurable FPUs

for executing the subsequent accurate or approximate software blocks without any problem in power retention. Further, this ensures that the error significance threshold is always met, but limits the use of the recovery mechanism to those cases where the error is present on the most significant bits. To characterize vulnerability of every FP pipeline to the timing error, we use FPV which is defined as the percentage of cycles in which a timing error occurs on the pipeline reported by the EDS sensors. To compute FPV, the ECU dynamically characterizes this per-pipeline metric over a programmable sampling period. The characterized FPV of each pipeline is visible to the software through the memory-mapped registers. Thus, the runtime scheduler leverages this characterized information for better utilization of FP pipelines, for example, it can assign fewer operations to a pipeline with higher FPV metadata. The runtime scheduler can also demote an error-prone pipeline to the approximate mode.

## 9.3.2 OpenMP Compiler Extension for Approximation

We provide two custom directives to OpenMP to identify approximate or accurate computations with an arbitrary granularity determined by the size of the structured block enclosed by the two custom directives:

```
#pragma omp accurate
structured-block

#pragma omp approximate [clause]
structured-block
```

The `approximate` directive allows the programmer to specify the tolerated error for the specific computation through an additional *clause*:

```
error_significance_threshold (<value N>)
```

The error is specified as the least significant *N* bits of the fraction. By default, if the programmer does not specify an error significance threshold, it is assumed zero-tolerance (i.e., the `approximate` directive behaves as the `accurate`). By using this clause the `approximate` structured blocks have deterministic fully predictive semantics: the maximum error significance for every FP instruction of the structured block is bound below the less significant *N* bits of the fraction. Moreover, any approximate instruction cannot modify any register other than its own. Let us consider the code snippet for Gaussian filter in Fig. 9.2.

Here, the programmer has indicated the whole `parallel` block as the accurate computation, with the exception of the FP multiplication and accumulation of the input data. These two operations are annotated for the approximate computation with a tolerance threshold of less significant 20 bits of the fraction derived from a profiling stage. We use a profiling technique [8] to identify tolerable error significance and

```
#pragma omp parallel
{
 #pragma omp accurate
 #pragma omp for
 for (i=K/2; i <(IMG_M-K/2); ++i) {
 // iterate over image
  for (j=K/2; j <(IMG_N-K/2); ++j) {
   float sum = 0;
   int ii, jj;
   for (ii =-K/2; ii<=K/2; ++ii) {
   // iterate over kernel
    for (jj = -K/2; jj <= K/2; ++jj) {
     float data = in[i+ii][j+jj];
     float coef = coeffs[ii+K/2][jj+K/2];
     float result;
     #pragma omp approximate \
              error_significance_threshold(20)
       {
        result = data * coef;
        sum += result;
       }
    }
   }
   out[i][j]=sum/scale;
  }
 }
}
```

**Fig. 9.2** Code snippet for Gaussian filter utilizing OpenMP approximation directives

error rate thresholds in error-tolerant image processing applications. The compiler transforms the blocks to appropriate API calls implemented through the runtime library.

### *9.3.3 Runtime Support*

The runtime library is a software layer that lies between the variation-tolerant shared-FPU architecture and the compiler-transformed OpenMP program. The goal of our runtime scheduler is to inspect the status of the FPUs and allocate them to `approximate` or `accurate` software blocks to reduce the overall cost of timing error correction. This is accomplished in a twofold manner: (i) the runtime scheduler reduces the number of recovery cycles for accurate blocks by favoring utilization of FPUs with a lower FPV, thus lower the error rate and energy; (ii) the scheduler further reduces the cost of error correction by deliberately propagating the error toward the program, thus excluding the correction cost. The latter guarantees the quality of service for `approximate` blocks by demoting FPUs to the approximate mode for

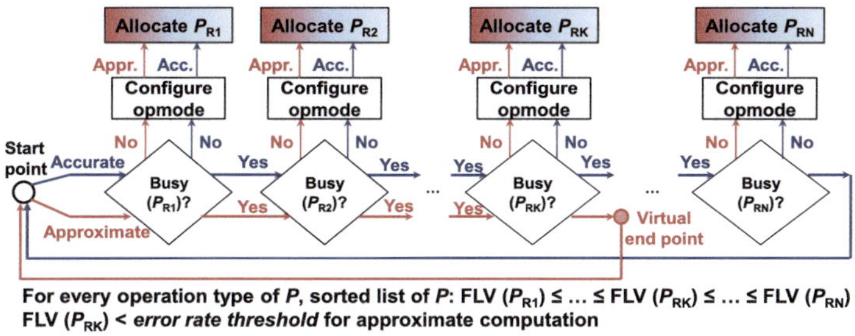


**Fig. 9.3** RANK scheduling based on FPV ranks

ignoring errors that match the tolerance expressed via the error significance threshold clause.

To allow for quick selection of best suited units for the accuracy target at hand, our scheduler ranks all the individual pipelines based on their FPV. The scheduler traverses the sorted list, starting from the head, until it finds an available pipeline. Once the target FP pipeline has been identified, it is configured to the desired operation mode on-the-fly, and a handler is returned to the program for offloading the consecutive FP instruction. Using this, for every type of FP operations the ranking algorithm tries to highly utilize those pipelines with a lower FPV (and rarely allocate operations to the pipelines at the end of list), thus the aggregate recovery cycles for execution of FP operations will be reduced. Figure 9.3 illustrates the ranking algorithm (RANK). For the approximate operations, in case of specifying an error rate threshold the scheduler limits its search to a certain element of the sorted list, e.g., until the *Kth* pipeline in Fig. 9.3.

As soon as the scheduler finds a pipeline which has a higher FPV than the error rate threshold, it marks it as the virtual end point of the list for the approximate operations. Therefore, for the following approximate requests, the scheduler starts from the start point of the sorted list, and traverses down toward the virtual end point of the corresponding sorted list for finding a free pipeline. However, this virtualization technique limits the available parallelisms discussed in the Sect. 9.4.

### *9.3.4 Application-Driven Hardware FPU Synthesis and Optimization*

In the earlier sections, we describe the three essential components of our variability-aware OpenMP environment: the language directive extensions, the compiler and runtime support, and the accuracy-configurable architecture. In this section, we introduce an optional yet effective methodology to generate efficient hardware FPU. The design flow should be done by choosing a threshold that is acceptable on a wide class of application, and if an application cannot tolerate this type of inaccuracy,

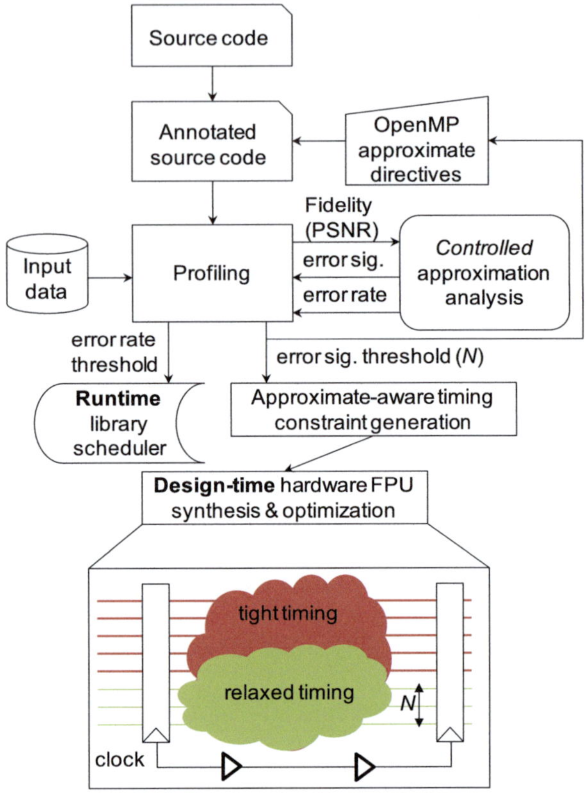


**Fig. 9.4** Methodology for application-driven hardware FPU synthesis and optimization

the runtime system must reconfigure architecture to the accurate mode. We couple the proposed methodology with the application tolerable error analysis presented in Sect. 9.2. As we have mentioned earlier, the output information of the profiling phase is two threshold values, i.e., the error significance threshold and the error rate threshold, that guarantee the acceptable fidelity metric (in our case: PSNR $\geq$30 dB). This information is utilized during design-time flow for synthesis and optimization of hardware FPUs; Fig. 9.4 illustrates the proposed methodology.

The error significance threshold indicates that any timing error below the bit position of, e.g., N can be ignored since it will not induce large deviations from the corrected value. This means for the approximate computation the only important parts are the bit positions higher than N since any timing error on these bits have to be corrected to guarantee the acceptable fidelity metric. Therefore, an efficient FPU for the approximate mode should eliminate the possibility of any timing error on the high-order bits, while relaxing this constraint on the low-order bits. At the same time they should not be too relaxed, to avoid the generation of many errors that have to be recovered in the accurate mode. Consequently, a set of tight timing constraints is generated to guide the hardware synthesis and optimization flow for providing fast paths connected to the high order bits (thus the lower delay, and the lower probability of timing errors). The synthesis CAD tool meets these constraints by utilizing fast leaky standard cells (low-$V_{TH}$) for the paths with the tight timing constraint, while utilizing the regular and slow standard cells (regular-$V_{TH}$ and high-$V_{TH}$) for the rest of paths. As a result, the new generated hardware FPU will experience a lower probability of the timing error on the bit positions higher than N, at the power expense of higher leaky cells.

We have applied the proposed methodology to optimize the netlist of the shared-FPUs. The approximation-aware timing constraints try to deliver fast paths connected to bit position of 20–32. As a result, the optimized shared-FPUs experience lower timing error rate; compared to the nonoptimized shared-FPUs, the total recovery cycles are reduced by 46 and 27% in the accurate and approximate modes, respectively. On the other hand, the total power over-head of the optimized shared-FPUs is 16% in comparison with the nonoptimized shared-FPUs (19% overhead in leakage power). However, this power overhead is highly compensated because the optimized shared-FPUs spend smaller number of clock cycles to compute the same amount of work. Experimental results in Sect. 9.4.1.3 quantify the energy benefit of this proposed methodology.

The proposed optimization methodology is based on either designer knowledge (provided from a domain expert), or static profiling (derived from the fidelity metric and error analysis). We should note that the static profiling is a common technique for approximate computation analysis [9, 10]. However, our methodology takes advantage of the maximum allowable error significance at design-time, while the error detection and correction circuits embedded in FPUs are responsible to dynamically handle any non-maskable timing error.

## 9.4 Experimental Results

We demonstrate our approach on an OpenMP-enabled SystemC-based virtual platform for on-chip multi-core shared-memory clusters with hardware accelerators [11]. Table 9.1 summarizes the architectural parameters. A cycle-accurate SystemC model of the shared-FPUs is also integrated to the virtual platform, which enables the variability-affected emulation. To accurately emulate the low-level device variability on the virtual platform, we have integrated the variability-induced error models at the level of individual FP pipelines using the instruction-level vulnerability characterization methodology presented in [12]. The RTL description of shared-FPUs are generated and optimized by FloPoCo [13], an arithmetic FP core generator of synthesizable VHDL. Then, the shared-FPUs have been synthesized for TSMC 45 nm technology, the general purpose process. The front-end flow with multi $V_{TH}$ cells has been performed using Synopsys Design Compiler with the topographical features, while Synopsys IC Compiler has been used for the back-end. The design has been typically optimized for timing to meet the signoff frequency of 625 MHz at (SS/0.81 V/125 °C).

**Table 9.1** Architectural parameters of shared-FPUs cluster

| | | | |
|---|---|---|---|
| ARM v6 core | 16 | TCDM banks | 16 |
| I$ size(per core) | 16 KB | TCDM latency | 2 cycles |
| I$ line | 4 words | TCDM size | 256 KB |
| Latency hit | 1 cycle | L3 latency | ≥60 cycles |
| Latency miss | ≥59 cycles | L3 size | 256 MB |
| Shared-FPUs | 8 | FP ADD latency | 2 |
| FP MUL latency | 2 | FP DIV latency | 18 |

Next, we have analyzed the delay variability of the shared-FPUs under process and temperature variations. First, to observe the effect of static process variation on the eight shared-FPUs, we have analyzed how the critical paths of each pipeline are affected due to within-die and die-to-die process parameters variation. Therefore, the various pipelines within the FPUs experience different variability-induced delay and thus display various error rate. During the sign-off stage, we have injected process variation in the shared-FPUs using the variation-aware timing analysis engine of Synopsys PrimeTime VX [14]. It utilizes process parameters and distributions of 45 nm variation-aware TSMC libraries [15] derived from first-level process parameters by principal component analysis. Second, to observe the effects of temperature variations, we employ voltage-temperature scaling feature of Synopsys PrimeTime to analyze the delay and power variations under temperature fluctuations. Finally, the variation-induced delay is back-annotated to the post-layout simulation to quantify the error rate of individual pipelines. For every back-annotated variation scenarios, the FP pipelines are characterized with a representative random set of $10^7$ inputs, automatically generated by FloPoCo. Finally, these error rate models are integrated to the corresponding modules in the SystemC virtual platform to emulate variability.

## 9.4.1 *Error-Tolerant Applications*

In this section, we evaluate the effectiveness of the proposed variability-aware OpenMP environment under the process variability for the error-tolerant image processing applications. For benchmark, we consider two widely used image processing applications as the approximate programs: Gaussian smoothing filter, and Sobel edge detection filter.

### 9.4.1.1 Execution Without Approximation Directives

For the first experiments, we marked the entire program for accurate computation (`#pragma omp accurate`), representative of what a nonexpert programmer would achieve without application profiling, tuning, and code annotation. Later,

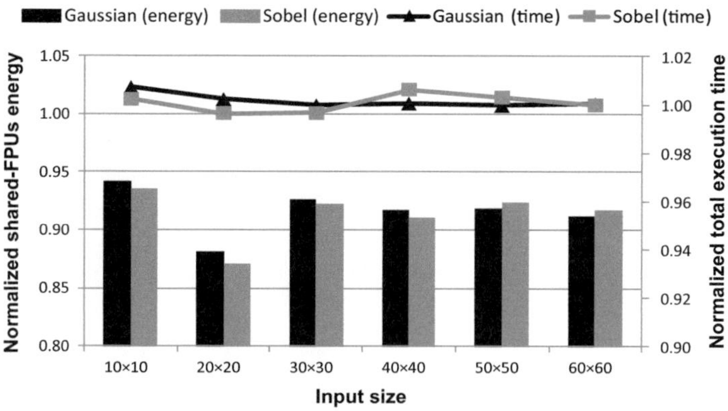


**Fig. 9.5** Energy and execution time of RANK scheduling (normalized to RR) for accurate Gaussian and Sobel filters

we show how these applications can benefit from the approximate code annotation. We have compared the proposed ranking scheduling (RANK) with the baseline round-robin scheduling (RR) in terms of FP energy and total execution time. The RR algorithm assigns the FP operations to the pipelines in the order they become available, while RANK utilizes the sorted list structure of the FPV. Figure 9.5 shows the shared-FPU energy and total execution time for the target applications for RANK normalized to the baseline RR algorithm. Each bar (or point) indicates the normalized shared-FPUs energy (or the total execution time) for a set of different input sizes.

As shown, the RANK algorithm achieves up to 12% lower energy for the shared-FPU compared to RR algorithm, while the maximum timing penalty is less than 1%. This energy saving is achieved by leveraging the characterized FPV metadata and the sorted list data structure that enable high utilization of those pipelines that display lower error rates. Consequently, it reduces the total recovery cycles, and energy. Moreover, the total timing overhead of the RANK is minimal, and the overhead for sorting and searching among eight shared-FPUs is highly amortized. These low cost features are accomplished through the advantages of fast TCDM, carefully placing the key data structures in TCDM, and the low-latency logarithmic interconnection.

#### 9.4.1.2 Profiling Error-Tolerant Applications

In this section, we present the profiling phases for producing useful threshold information to enable approximate computation. We analyze the manifestation of a range of error significance and error rate on the PSNR of the two image processing applications. We have annotated the approximable regions of the application codes using the proposed OpenMP custom directives (the code snippet for the Gaussian filter is shown in Fig. 9.2). The annotated approximate regions of both applications are only composed of FP addition and multiplication operations. We quantify how much error

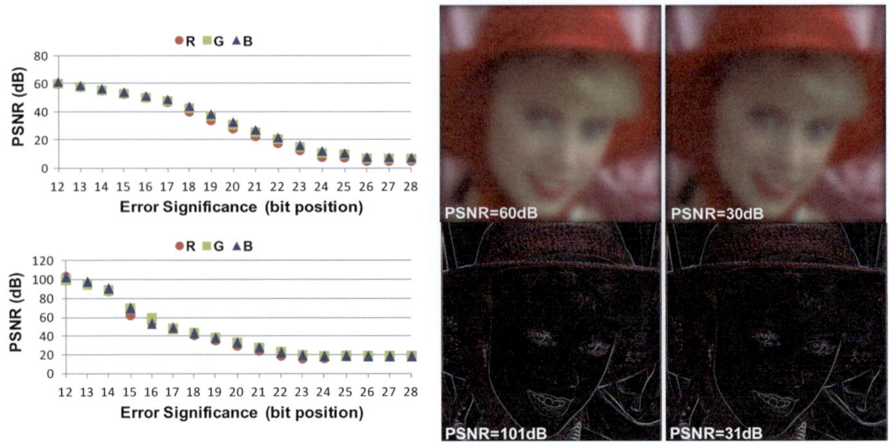


**Fig. 9.6** PSNR degradation as a function of error significance: **a** for Gaussian filter (*top*); **b** for Sobel filter (*bottom*)

significance can be tolerated in these approximate regions, given a maximum error rate. To do so, we have profiled the annotated approximate regions of the programs. In a series of profiling, we have monotonically increased the error significance by injecting the timing errors as random multiple-bit toggling up to a certain bit position of the FP single precision representation. The position of multiple-bit toggling is varied from 1 to 28, for a wide range of 1% error rate to 100%.

Figure 9.6 illustrates results for the error rate of 100%, i.e., every addition and multiplication operation in the FP approximate regions has an errant output depending up on the injected error significance. Figure 9.6a shows the PSNR degradation of output image of the Gaussian filter as a function of the error significance. As shown, the three channels of RGB color space, experience similar PSNR degradations by increasing the error significance. Figure 9.6b also illustrates the similar trend for the Sobel filter. The rightmost part of Fig. 9.6 shows that this degradation of the quality is acceptable from the user's perspective. In summary, the output information of these profiling indicates that for a given error rate of 100, 50, 25% if the timing error lies within the bit position of 0–20, 21, 22 of the fraction part, these two applications can tolerate the timing error by delivering a PSNR of greater than 30 dB. This information is essential not only during runtime to intentionally ignore the tolerable timing errors, but also for efficient hardware FPU synthesis and optimizations, detailed in the following section.

Therefore, for the approximate regions of these applications, we have set the error rate threshold to 100%, and the error significance threshold to 20 to maintain the acceptable PSNR. By setting the threshold of the error rate to 100%, during the runtime execution of the approximate regions all FPUs can be utilized. This is important in data-parallelized image applications where there is enough parallelism, and especially so when the number of FPUs is lower than the number of the cores and any time-multiplexing might incur performance degradation.

#### 9.4.1.3 Execution with Approximation Directives

Now, let us quantify the benefit of the approximate computation using the information of the profiling. Since the RANK scheduling algorithm surpasses the baseline RR algorithm, for the rest of results we have used the RANK algorithm. We have repeated the experiments in Sect. 9.4.1.1, but for two variants of the applications code. In the first version, the programs are entirely composed of the accurate FP operations, and the in the second version the programs utilize the approximate ADD and MUL operations in the annotated regions of code.

Figure 9.7 shows the total shared-FPUs energy for these two versions of the programs with different input sizes. The first group of bars shows the energy of the shared-FPUs for the accurate programs, while the second group of bars refers to the approximate programs. For example, with an input size of $60 \times 60$, the shared-FPUs consume 3.5 μJ (or 4.6 μJ) for the accurate Gaussian (or Sobel) program, while execution of the approximate version of the program reduces the energy to 2.8 μJ (or 3.5 μJ), achieving 24% (or 30%) energy saving. This energy saving is achieved by ignoring the timing error within the bit position of 0–20 of the fraction part. The next two bars show the energy of an optimized hardware implementation of the shared-FPUs, discussed in the following.

To generate the efficient FPUs suitable for these applications we leveraged the hardware FPU synthesis and optimization methodology proposed in Sect. 9.3.4. Therefore, the application-driven timing constraints guide the CAD flow to selectively optimize timing of the desired paths. Figure 9.7 also shows the energy differences between the nonoptimized and optimized FPUs in the two operating modes. On average, compared to the nonoptimized shared-FPUs, the optimized shared-FPUs achieves 25 and 7% lower energy for the accurate and approximate modes, respectively. Overall, utilization of the annotated programs with the approximate directives on top of the optimized shared-FPUs achieves an average energy saving of 36%.

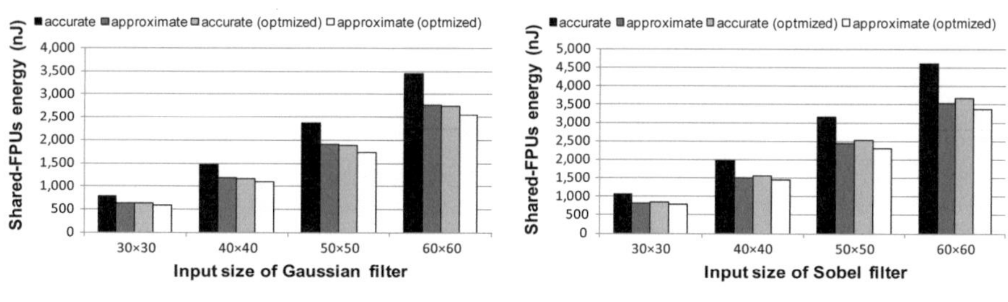


**Fig. 9.7** FP energy of accurate and approximate programs for nonoptimized and optimized hardware shared-FPUs

### 9.4.2 Error-Intolerant Applications

Using the concept of configurable accuracy as discussed earlier, we now show that the proposed variability-aware OpenMP environment not only facilitates efficient execution of the approximate programs, but also reduces the cost of recovery for the error-intolerant general-purpose applications. We have evaluated the effectiveness of our proposed approach in the presence of process variability under operating temperature fluctuations for five applications where we have no domain expert knowledge about their tolerance to error: three widely used 2-D computational kernels (matrix multiplication, matrix addition with scalar multiplication, and DCT), Monte Carlo kernel, and image conversion kernel (HSV2RGB).

Figure 9.8 shows the shared-FPUs energy saving of these applications compared to the worst-case design. For these experiments, we consider 25% voltage overdesigned for the baseline FPUs which can guarantee their error-free operations [4]. On average 22% (and up to 28%) energy saving is achieved at the operating temperature of 125 °C, thanks to allocating the FP operations to the appropriate pipelines. As shown, this saving is consistent (20–22% on average) across a wide temperature range ($\Delta$T=125°C), thanks to the online FPV metadata characterization which reflects the latest variations, thus enabling the scheduler to react accordingly. The lower temperature leads to a higher delay in the low-voltage region of nanometer CMOS technologies [16], thus the higher error rate and the more energy for recovery. Please note that after having the ranked pipelines tables on TCDM, we rarely need to re-execute the sorting algorithm unless we sense a temperature fluctuation which has a slow timing-constant.

We also compare our proposed environment with method presented in Truffle [4]. Truffle, as a single core architecture, duplicates all the functional units in the execution stage. Half of them are hardwired to $Vdd_{High}$ (to execute the accurate operations), while the other half operate at $Vdd_{Low}$ (to execute the approximate operations). To have an iso-area comparison with Truffle, as it is suggested in their paper, we assume that Truffle uses dual-voltage FPUs and changes the voltage depending on the instruction being executed. This would also save the static power. To have a fair comparison, we also assume that Truffle employs a fast Vdd-hopping technique

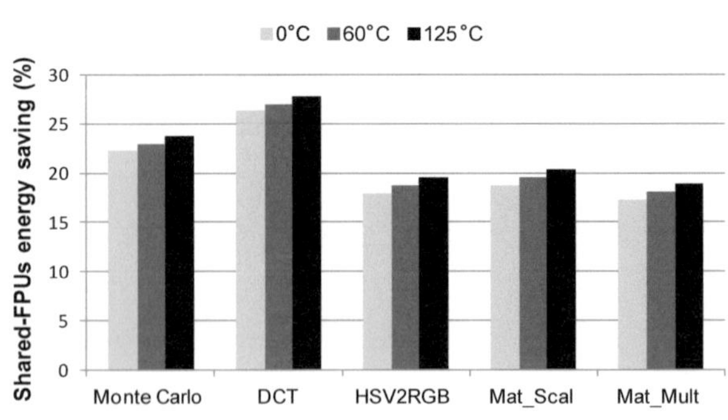


**Fig. 9.8** Shared-FPUs energy saving for the error-intolerant applica-tions compared to the worst-case design

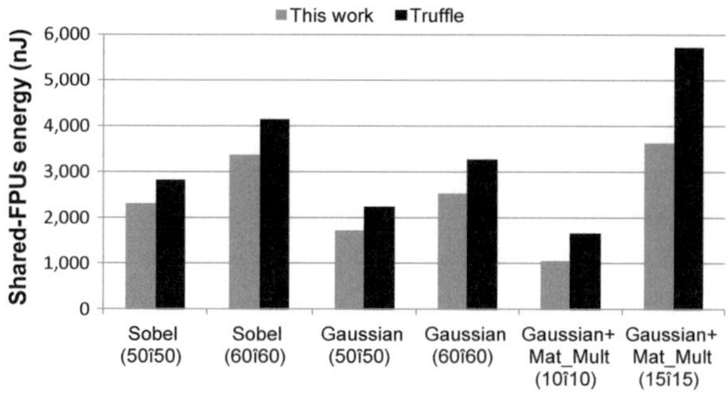


**Fig. 9.9** Energy comparison with Truffle: (i) only approximate; (ii) concurrent approximate and accurate applications

to switch between $Vdd_{High}$ and $Vdd_{Low}$. Among the Vdd-hopping implementation techniques [17], Beigne et al. propose a Vdd-hopping unit with voltage transitions less than 100 ns [17, 18]. Kim et al. also propose fast on-chip voltage regulators with transitions time of 15–20 ns [18], thus we consider this transition time and optimistically augment a latency of 10-cycle for switching FPUs between the accurate and approximate modes. We apply Truffle limitation to our virtual plat-form cluster to quantify its energy.

For comparison, we consider two application scenarios: (i) once the cluster is executing only one approximate application; (ii) simultaneous execution of one approximate application with one accurate application. In the former scenario, entire 16 cores of the cluster cooperatively execute one of the approximate image applications, while in the latter scenario, eight cores execute the approximate Gaussian filter and the other eight core execute the accurate matrix multiplication, simultaneously. Figure 9.9 compares the shared-FPUs energy of Truffle with our proposed approach when executing the above two scenarios. As shown, our proposed approach surpasses Truffle in the both applications scenarios. In the former scenario, on average, our approach saves 20% more energy compared to Truffle by reducing the conservative voltage overdesigned for the accurate part of filters application. For the mixed scenario of the applications, our approach saves 36% more energy, since Truffle highly faces with the overhead of frequent switching between the accurate and approximate modes which is imposed by interference of the accurate and approximate operations resulting from the concurrent execution of Gaussian and matrix multiplication applications.

## 9.5 Chapter Summary

We propose an OpenMP programming environment that is resilient to variability-induced timing errors and suitable for fine-grained interleaved approximate and accurate computation on shared-FPUs processor clusters. This is orchestrated through a vertical abstraction of circuit-level variations into a high-level parallel software execution. The OpenMP extensions help a programmer specify accurate and approximate FP parts of a program. The underlying architecture features a set of shared-FPUs with two sensing and actuation primitives; every FPU dynamically senses the timing errors, characterizes its own FPV metadata, and can be configured to operate in the approximate or accurate modes. The runtime scheduler utilizes the sensed FPV metadata, and parsimoniously actuates depending upon the code region requirements on the computational accuracy. These three components in the proposed environment support a controlled approximation computation through various design-time phases (applications profiling, and FPU synthesis and optimization) in combination with runtime sensing and actuation primitives. Either the environment deliberately ignores the otherwise expensive timing error correction in a fully controlled manner, or it tries to reduce the frequency of timing errors.

For general-purpose error-intolerant applications with no domain expert knowledge, our approach reduces energy up to 28%, across a wide temperature range ($\Delta$T=125°C), compared to the worst-case design. For error-tolerant image processing applications with the annotated approximate directives, on average, 36% energy saving is achieved while maintaining the PSNR $\geq$30 dB. In comparison with the state-of-the-art architecture [4], our approach saves 36% more energy when executing finely interleaved mixture of FP operations.

## References

1. M.R. Kakoee, I. Loi, L. Benini, Variation-tolerant architecture for ultra low power shared-l1 processor clusters. IEEE Trans. Circuits Syst. II Express Br. **59**(12), 927–931 (2012)
2. D. Jeon, M. Seok, Z. Zhang, D. Blaauw, D. Sylvester, Design methodology for voltage-overscaled ultra-low-power systems. IEEE Trans. Circuits Syst. II Express Br. **59**(12), 952–956 (2012)
3. M.A. Breuer, Intelligible test techniques to support error-tolerance, in *13th Asian Test Symposium (ATS 2004), 15–17 November 2004* (Kenting, Taiwan, 2004), pp. 386–393
4. H. Esmaeilzadeh, A. Sampson, L. Ceze, D. Burger, Architecture support for disciplined approximate programming, in *Proceedings of the Seventeenth International Conference on Architectural Support for Programming Languages and Operating Systems, ASPLOS XVII* (ACM, New York, NY, USA, 2012), pp. 301–312
5. H. Cho, L. Leem, S. Mitra, Ersa: error resilient system architecture for probabilistic applications. IEEE Trans. Comput. Aided Des. Int. Circuits Syst. **31**(4), 546–558 (2012)
6. A. Sampson, W. Dietl, E. Fortuna, D. Gnanapragasam, L. Ceze, D. Grossman, Enerj: approximate data types for safe and general low-power computation, in *Proceedings of the 32nd ACM SIGPLAN Conference on Programming Language Design and Implementation, PLDI '11* (ACM, New York, NY, USA, 2011), pp. 164–174
7. K.A. Bowman, J.W. Tschanz, S.L. Lu, P.A. Aseron, M.M. Khellah, A. Raychowdhury, B.M. Geuskens, C. Tokunaga, C.B. Wilkerson, T. Karnik, V.K. De, A 45 nm resilient microprocessor core for dynamic variation tolerance. IEEE J. Sol. State Circuits **46**(1), 194–208 (2011)
8. A. Rahimi, A. Marongiu, R.K. Gupta, L. Benini, A variability-aware openMP environment for efficient execution of accuracy-configurable computation on shared-FPU processor clusters, in *2013 International Conference on Hardware/Software Codesign and System Synthesis (CODES+ISSS)* (2013), pp. 1–10
9. W. Baek, T.M. Chilimbi, Green: a framework for supporting energy-conscious programming using controlled approximation, in *Proceedings of the 2010 ACM SIGPLAN Conference on*

*Programming Language Design and Implementation, PLDI '10* (ACM, New York, NY, USA, 2010), pp. 198–209
10. M.S.K. Lau, K.-V. Ling, Y.-C. Chu, Energy-aware probabilistic multiplier: Design and analysis, in *Proceedings of the 2009 International Conference on Compilers, Architecture, and Synthesis for Embedded Systems, CASES '09* (ACM, New York, NY, USA, 2009), pp. 281–290
11. P. Burgio, A. Marongiu, D. Heller, C. Chavet, P. Coussy, L. Benini, Openmp-based synergistic parallelization and HW acceleration for on-chip shared-memory clusters, in *15th Euromicro Conference on Digital System Design, DSD 2012, Cesme, Izmir, Turkey, September 5–8* (2012), pp. 751–758
12. A. Rahimi, L. Benini, R.K. Gupta, Analysis of instruction-level vulnerability to dynamic voltage and temperature variations, in *Design, Automation Test in Europe Conference Exhibition (DATE)* (2012), pp. 1102–1105
13. Flopoco: Floating-point Cores Generator. http://flopoco.gforge.inria.fr/
14. PrimeTime VX User Guide, June 2011
15. TSMC 45 nm Standard Cell Library Release Note, v 120a, November 2009
16. R. Kumar, V. Kursun, Reversed temperature-dependent propagation delay characteristics in nanometer cmos circuits. IEEE Trans. Circuits Syst. II Express Br. **53**(10), 1078–1082 (2006)
17. E. Beigne, F. Clermidy, H. Lhermet, S. Miermont, Y. Thonnart, X.-T. Tran, A. Valentian, D. Varreau, P. Vivet, X. Popon, H. Lebreton, An asynchronous power aware and adaptive noc based circuit. IEEE J. Sol. State Circuits **44**(4), 1167–1177 (2009)
18. W. Kim, D.M. Brooks, Gu-Y. Wei, A fully-integrated 3-level DC/DC converter for nanosecond-scale DVS with fast shunt regulation, in *IEEE International Solid-State Circuits Conference, ISSCC 2011, Digest of Technical Papers, San Francisco, CA, USA, 20–24 February* (2011), pp. 268–270

# Chapter 10
# An Approximation Workflow for Exploiting Data-Level Parallelism in FPGA Acceleration

**Abstract** Modern applications including graphics, multimedia, web search, and data analytics not only can benefit from acceleration, but also exhibit significant degrees of tolerance to imprecise computation. This amenability to approximation provides an opportunity to trade quality of the results for higher performance and better resource utilization. Exploiting this opportunity is particularly important for FPGA accelerators that are inherently subject to many resource constraints. To better utilize the FPGA resources, we devise, GRATER, an automated design workflow for FPGA accelerators that leverages imprecise computation to increase data-level parallelism and achieve higher computational throughput. The core of our workflow is a source-to-source compiler that takes in an input kernel and applies a novel optimization technique that selectively reduces the precision of kernel's data and operations. By selectively reducing the precision of the data and operation, the required area to synthesize the kernels on the FPGA decreases allowing to integrate a larger number of operations and *parallel* kernels in the fixed area of the FPGA. The larger number of integrated kernels provides more hardware context to better exploit data-level parallelism in the target applications. To effectively explore the possible design space of approximate kernels, we exploit a genetic algorithm to find a subset of safe-to-approximate operations and data elements and then tune their precision levels until the desired output quality is achieved. GRATER exploits a fully software technique and does not require any changes to the underlying FPGA hardware. We evaluate GRATER on a diverse set of data-intensive OpenCL benchmarks from the AMD SDK. The synthesis result on a modern Altera FPGA shows that our approximation workflow yields 1.4×–3.0× higher throughput with less than 1% quality loss.

## 10.1 Introduction

Before the effective end of Dennard scaling, we were able to improve all three of performance, efficiency, and generality. With the end of Dennard scaling, the community is facing an iron triangle. We can only improve any two of the performance, efficiency, and generality at the expense of the third. Solutions that provide significant performance and efficiency gains while retaining as much generality as possible, are



A. Rahimi et al., *From Variability Tolerance to Approximate Computing in Parallel Integrated Architectures and Accelerators*,
DOI 10.1007/978-3-319-53768-9_10

highly desirable. One promising approach is the use of programmable accelerators, such as FPGAs to achieve higher performance and efficiency in certain application domains. Approximate computing is another promising approach that leverages tolerance of applications to imprecision and trades small losses of quality for gains in performance and efficiency [1–5].

The confluence of these two trends can potentially yield significant performance and efficiency gains since many of the applications that can benefit from the FPGA accelerations are amenable to approximation. However, there is a lack of techniques that exploits this opportunity. This work aims to bridge the gap between approximation and the FPGA acceleration through an automated design workflow. Altera and Xilinx recently offer high-level acceleration frameworks for OpenCL [6, 7], hence in our work we target acceleration of data-intensive computational OpenCL applications. The challenge is, however, devising a workflow that can be plugged into the existing toolsets and can automatically identify the opportunities for approximation while keeping the quality loss reasonably low. This chapter addresses this challenge and makes the following contributions:

1. We propose GRATER, a design workflow that automatically leverages approximation to provide more opportunities for the FPGA accelerators to utilize data-level parallelism and achieve higher throughput. GRATER automatically reduces required hardware area for synthesizing an instance of an OpenCL kernel. This required area is determined by the precision of data and operations which is specified by the kernel program. By selectively reducing the precision, a larger number of *parallel* approximate kernels can be mapped in the fixed area budget of an FPGA. GRATER provides a readily applicable workflow that exploits the inherent error tolerance of the emerging applications for higher computational throughput with off-the-shelf FPGAs without any changes to their hardware structure.
2. GRATER systematically tunes the precision of the operations and data in the input OpenCL kernel, subject to a statistical target for quality-of-result. GRATER uses a source-to-source compiler that leverages an automated transformation to selectively reduce the precision. The precision of the data and operations are automatically inferred from the precision of their operands. We devise a genetic programming-based optimization algorithm that assigns various precision levels to different data and operations in the kernel. We use genetic programming to evolve kernel variants until one is found with optimal assignments that reduces synthesized kernel area while stochastically satisfying the quality-of-result target.
3. We evaluate GRATER with a diverse set of data-intensive OpenCL benchmarks selected from the AMD APP SDK v2.9 [8]. The synthesis result on an Altera Stratix V FPGA shows that the reduced area of the transformed approximate kernels yields 1.4×–3.0× higher throughput with less than 1% loss of quality.

The rest of the chapter is organized as follows. Section 10.2 describes acceleration of OpenCL applications on the FPGAs. GRATER approximation design workflow is presented in Sect. 10.3. In Sect. 10.4, we present experimental results, followed by conclusion in Sect. 10.5.

## 10.2 OpenCL Execution Model

OpenCL is a platform-independent framework for writing programs that execute across a heterogeneous system consisting of multiple compute devices including CPUs or accelerators such as GPUs, DSPs, and FPGAs. OpenCL uses a subset of ISO C99 with added extensions for supporting data and task-based parallel programming models. The programming model in OpenCL comprises of one or more device kernel codes in tandem with the host code. The host code typically runs on a CPU and launches kernels on other compute devices like the GPUs, DSPs, and/or FPGAs through API calls. The instance of an OpenCL kernel is called a work-item. These kernels execute on compute devices that are a set of compute units (CUs), each comprising of multiple processing elements having ALUs. The work-items execute on a single processing element and exercise the ALU. The OpenCL platform model from the programming model to the framework of the compute devices is illustrated in Fig. 10.1.

### *10.2.1 Mapping OpenCL Programs on FPGAs*

The Altera OpenCL SDK [6] allows programmers to use high-level OpenCL kernels, written for GPUs, to generate an FPGA design with higher performance per Watt [9]. An OpenCL kernel is first compiled and then synthesized as a special dedicated hardware for mapping on an FPGA. However, GPUs and FPGAs exploit data-level parallelism differently, which leads to disparate benefit in terms of performance per

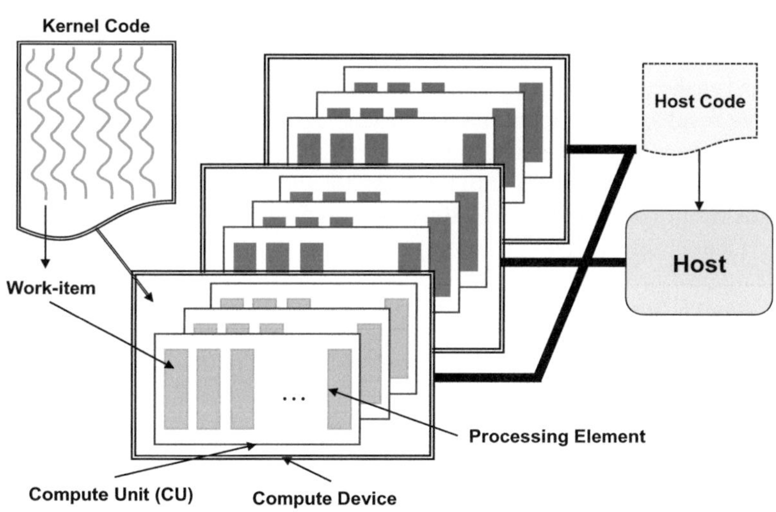


**Fig. 10.1** OpenCL platform model

watt. GPUs are single-instruction multiple-data (SIMD) devices that exploit data-level parallelism: they group processing elements in a CU to perform the same operation but on their own individual data. On the other hand, FPGAs exploit pipeline parallelism in a CU where different stages of the instructions are applied to different work-items concurrently leading to a higher performance per Watt.

FPGAs can further improve the performance benefits by creating multiple copies of the kernel pipelines (synthesized version of an OpenCL kernel).[1] For instance, this replication process can make $N$ copies of the kernel pipeline. As the kernel pipelines can be executed independently from one another, the performance would scale linearly with the number of copies created owing to the data-level parallelism model supported by OpenCL. In the following sections, we describe how GRATER can reduce the amount of resources for a kernel pipeline to save area and exploit remaining area resources to boost performance by replication. GRATER systematically reduces the precision of data and operations in OpenCL kernels to shrink the resources used per kernel pipeline by transforming complex kernels to simple kernels that produce *approximate* results.

## 10.3 GRATER: Approximation Design Workflow

GRATER supports a source-to-source compiler to generate approximate kernels via source-to-source OpenCL kernel transformation. The transformation algorithm automatically detects and simplifies parts of the kernel code that can be executed with reduced precision while preserving the desired quality-of-result. To achieve this goal, GRATER takes in as inputs, an *exact* OpenCL kernel, a set of input test cases, and a metric for measuring the quality-of-result target. GRATER compiler investigates the exact kernel code and detects data elements, i.e., OpenCL kernel variables, that provide possible opportunities for increased performance in exchange of accuracy. GRATER then automatically generates a set of approximate kernels that produce acceptable results. These approximate kernels provide improved performance benefits by reducing the area when implemented on the FPGAs. GRATER outputs an optimized approximate kernel with the least area whose output quality satisfies the quality-of-result target. Figure 10.2 illustrates an overview of our workflow.

GRATER uses the precision of the operations and data to tune performance as a tradeoff against precision. The transformation investigates a set of kernels where in each version, some of these potential variables are replaced with a less accurate variable.[2] We assign a precision tag (PT) to each variable type. For example, a kernel with data types ranging from floating point to char has four levels of complexity:

---

[1]Replication is handled in Altera OpenCL by setting `num_compute_units` as a kernel attribute.

[2]We limit the space of our optimization search across the available variable types in OpenCL, as opposed to within a type itself [10], due to the nature of a source-to-source transformer that requires to work at the same level of abstraction of the input programming language. GRATER enables Altera OpenCL synthesis tool chain to benefit from this source-to-source translation by generating standard OpenCL approximate kernels.

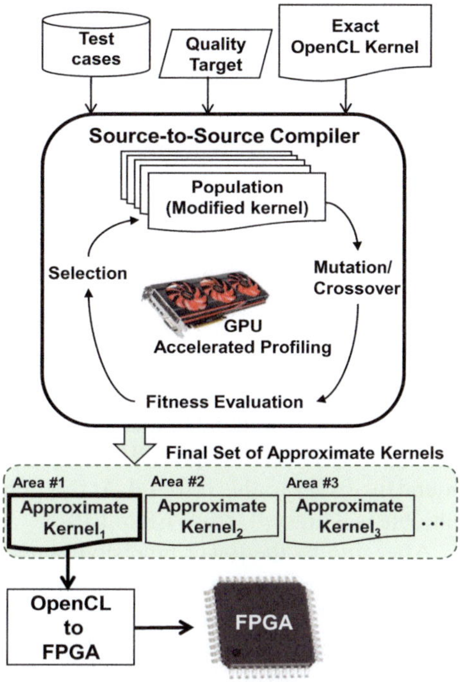


**Fig. 10.2** Overview of GRATER, our approximation design workflow

{4, 3, 2, 1} are assigned to {`float`, `int`, `short`, `char`} respectively. The higher the PT, the higher the accuracy requirements, and the higher resource consumption. A brute-force methodology for exploring the approximate kernels is to generate an approximate kernel for every possible combination of the variable types. For instance, for a kernel with |V| number of `float` variables, a total number of $4^{|V|}$ kernels would be generated where in each version every `float` variable is replaced by different PTs. This results in an exponentially growing design space intractable to search. To avoid this huge design space exploration, we devise an algorithm that first detects those variables that are amenable to approximation and then applies a genetic-based algorithm to approximate the kernel. We discuss the details of our algorithm in the following subsections.

### *10.3.1 Analysis and Pruning*

In the first step, GRATER detects variables in the code that are amenable to approximation. To do so, a separate kernel is generated for testing the amenability of every variable. In each kernel, the precision of one variable is demoted by one level (a $\Delta$PT demoting), while other variables have their exact precision, to measure the *significance* of a small precision loss of a variable on the quality of result. This test determines whether the precision of the selected variable can be reduced or not. If the output quality is less than the desired output quality, GRATER excludes this variable from the set of safe-to-approximate variables and does not modify its precision.[3] Consequently, the variable is eliminated from the candidate list of variables for approximation. The pruning algorithm continues the screening process for all the variables in the code (Line 4–10 in Algorithm 1). The pruning algorithm is executed |V| times to determine approximable variables (AV). This sensitivity test is done with the help of profiling feedback that is accelerated on a GPU.

GRATER then finds the lowest possible precision for each variable in AV (Line 11–14 in Algorithm 1). It generates an approximate kernel for every variable in AV, where in each kernel, one variable type is replaced by the lowest possible precision (e.g., `char`) while other variables preserve their exact precision (EP) that originally have in the exact code. If the quality of the generated kernel is less than the desired output quality, then that tentative lowest precision is promoted by one level and the same quality check is repeated. This process is continued until the lower precision bound for each variable is found. At this point, PT value ranges for each approximable variable is extracted (from EP to LP).

After finding the lower precision bound for all variables in AV, another approximate kernel is generated in which all approximable variables get their lowest possible precision (LP values) found in the previous step. If this kernel meets the quality-of-result target, the solution is found (Line 15–19 in Algorithm 1). Otherwise, a genetic algorithm, described in the following, is run to find the approximate kernel.

## *10.3.2 Genetic-Based Approximation Algorithm*

Genetic algorithm is a powerful stochastic search method which is deployed to find a good solution from a large search space [11]. To operate with a genetic algorithm, we need to take into account the following components: (1) a genetic representation of solutions in a form that can be interpreted as a chromosome, (2) an initial population, (3) a fitness function which gives an evaluation of the desirability of each chromosome, and (4) genetic operators that change the composition of new generation during reproduction and a selection operator for choosing the survivors.

### 10.3.2.1 Genetic Representation of Chromosomes

We represent each individual as an array of precision tags with the length of AV list. Each gene in this representation shows the PT of each variable in AV. Every individual can be easily translated to a candidate approximate kernel. The precision

[3]GRATER also enables the programmer to annotate critical variables as non-approximable, so that the transcompiler would not change their precision.

**Algorithm 1** Pseudo-code for the GRATER

```
1: function PRUNE&RELAX(ExactKernel, QualityTarget, inputSet)
2:     V = {All candidate variables in ExactKernel}     AV = {}
3:     TopPop = {}     cInput = input_0 from inputSet
4:     for all variables v_i in V do
5:         generate kernel_i s.t. v_i ← ΔPT demoting
6:         run kernel_i with cInput on GPU
7:         if (Quality(kernel_i) ≥ QualityTarget) then
8:             AV = AV ∪ v_i
9:         end if
10:    end for
11:    for all variables v_i in AV do
12:        LP_i = FindLowerPT(v_i, cInput)
13:        EP_i = getExactPT(v_i, cInput)
14:    end for
15:    generate kernel_min s.t. ∀ v_i ← LP_i
16:    run kernel_min with cInput on GPU
17:    if (Quality (kernel_min) ≥ QualityTarget) then
18:        ApproxKernel = kernel_min
19:    else
20:        ApproxKernel, TopPop =
21:            GA(ExactKernel, LP, EP, cInput)
22:    end if
23:    for all input_i in the training inputSet do
24:        run ApproxKernel with input_i on GPU
25:        if (Quality(ApproxKernel) < QualityTarget) then
26:            NeedToChangeSolution = True
27:            for all kernel_j in TopPop do
28:                run kernel_j with input_i on GPU
29:                if (Quality(kernel_j) > QualityTarget) then
30:                    ApproxKernel = kernel_j
31:                    NeedToChangeSolution = False
32:                    Break
33:                end if
34:            end for
35:            if (NeedToChangeSolution) then
36:                cInput = input_i
37:                Goto line 11
38:            end if
39:        end if
40:    end for
41:    return ApproxKernel
42: end function
```

of the variables and associated operations in the approximate kernel is inferred from the assigned PT value in the chromosome.

#### 10.3.2.2 Population

The initial population is randomly generated. Each approximable variable can have a PT value range with different levels of complexity, started from its lowest precision bound to its exact precision level (LP and EP in Algorithm 1).

All individuals in the population should meet the desired quality-of-result requirement. This can be verified either by executing the kernel or comparing its PT values with the least precision chromosome found. The least precision chromosome found in the population is the one that the PT values of every gene in its chromosome is lower than the PT values of corresponding genes in all other chromosomes. If such a chromosome does not exist in the population, the least precision PT in the population would be the same as LP. In this case, for all generated kernels we need a kernel execution for accuracy check. When the quality measurement test is done by executing an approximate kernel, its output is compared with the exact kernel output on a representative data input. If the output of the approximate kernel cannot satisfy the quality-of-result target, the approximate kernel is ruled out from the population. Otherwise, it is considered as one of the candidates for the next generation. This kernel profiling and execution process is accelerated on a GPU. This is accomplished by decoupling the quality loss analysis and the approximate kernel mapping thanks to the platform-independent nature of OpenCL. To increase the speed of genetic algorithm, before creating and executing each approximate kernel, the generated chromosome is compared to the chromosome with the least PT in the population so far. If all PT values in the newly generated chromosome is higher than or equal to the PT values of the least precision chromosome, this new chromosome can certainly meet the quality-of-result target; otherwise, the corresponding kernel should be executed for accuracy check.

#### 10.3.2.3 Fitness Function

Given a kernel, the fitness function returns a value showing the desirability of the approximate kernel. The fitness value is used by the selection operation to decide which individuals would survive to the next generation. Our main objective is to find an approximate kernel that minimizes the resource utilization on FPGA while meeting the quality-of-result requirement. To achieve this objective, our fitness function computes a weighted summation of its assigned PT values in the chromosome to estimate the area occupancy. For each variable, the weight is determined by a coefficient assigned to each precision tag multiplied by the number of times the variable is used in operations in the kernel. (The coefficients are determined through simulations which is 0, 1, 2, 6 for PT of 1, 2, 3, 4 respectively.) The higher the precision and the number of times the variable is used in operations, the higher weight it gets. With this definition, the lower the fitness value, the lower area occupancy that configuration has.

#### 10.3.2.4 Selection and Genetic Operators

We use two genetic operators, crossover and mutation, to produce new chromosomes. Crossover combines the first part from one parent chromosome to the second part from the other parent chromosome to produce a child chromosome. In this implementation, the crossover point is selected randomly. Mutation operation randomly modifies PT values of approximable variables in the chromosome. The new PT value is a random value in the range of LP and EP for the approximable variable. The newly generated chromosome is only accepted if it meets the quality-of-result requirement; otherwise, the operation is applied again.

There are many possible selection algorithms to select more fit individuals from the new and old population for the next generation. To rank area occupancy of the approximate kernels without synthesizing and mapping the kernel on FPGA, we use the fitness values as an estimate of the area occupancy. The selected chromosomes are sorted based on their estimation of area occupancy (fitness value) in each iteration. The top best individuals are always transferred for the next generation (elitism selection). For the rest, individuals are selected based on the proportionate selection where some of them might change with the crossover and mutation operations. For the simulation purpose, the crossover rate, mutation rate, and elitism rate is 0.7, 0.05, and 0.25 respectively. The algorithm runs as long as the user defined number of iterations has not been passed yet or when the best fitness values stop growing any further.

Until here, the genetic algorithm finds the final solution using only one input test case. This solution is verified with the other input test cases from the training set. If it meets the quality-of-result requirement for all inputs of this set, this approximate kernel is the final solution. Otherwise, either the other top chromosomes in the population is checked or the genetic algorithm is applied again for the failed input (Line 23–40 in Algorithm 1).

When this procedure is terminated, the best chromosome with the lowest fitness value is selected and translated to its corresponding approximate kernel which has the least area estimation on FPGA. This kernel is passed to the Altera SDK tool to be synthesized and mapped on the FPGA.

## 10.4 Experimental Results

### *10.4.1 Experimental Setup*

We focus on a diverse set of application domains, including image processing (recursive gaussian, sobel), signal processing (convolution, dct), and physical simulation (n-body). These benchmarks are selected from the AMD accelerated parallel processing (APP) SDK v2.9 [8]; that is, a complete development platform created by the AMD to leverage accelerated compute using OpenCL. All of these applications are error tolerant and have approximable data in their kernels. The number of variables in these kernels are in the range between 11 and 17.

GRATER source-to-source compiler [12] is implemented in Python, and accepts the exact kernel, the desired quality metric, and a set of 100 training input test cases as its inputs. GRATER utilizes the AMD Evergreen Radeon HD 5870 GPU device for accelerated profiling experiments, and finally generates an optimized approximate kernel with the minimum area occupancy estimation and an acceptable output error. The approximate OpenCL kernels were synthesized for Altera DE5 board with a

Stratix V FPGA using Altera OpenCL SDK 13.1 tool [6].[4] Sections 10.4.2 and 10.4.3 detail how GRATER can reduce the area and correspondingly increase the throughput for different applications.

### *10.4.2 Area Savings with Approximate Kernels*

Table 10.1 shows the resource utilization for an exact kernel and the optimized approximate kernel. As shown, the area utilization is reduced by an average of 14–25% for different FPGA resources using the transformed approximate kernel instead of the exact one. This is achieved by the proper precision tuning of the kernel variable types that brings area saving without scarifying the output quality. Figure 10.3 compares the maximum number of mapped kernels on the FPGA board for the exact kernels and the approximate kernels. For instance, the exact r-gaussian kernel contains 12 `float` variables in which 5 of them are converted to `short` and one of them to `char`. This precision tuning reduces DSP block utilization by 45% and logic utilization by 11%, hence enables mapping 9 approximate kernels rather than 5 exact kernels. Considering the geometric mean across all the benchmarks, GRATER is able to map 12 approximate kernels instead of 6 exact kernels in the fixed area budget of the FPGA. Given a fixed area budget of the FPGA, GRATER improves the number of mapped kernel by a factor of 2× on average (maximum 3.3× in n-body).

### *10.4.3 Speedup*

As shown in Sect. 10.4.2, GRATER reduces the synthesized area for the approximate kernels on the FPGA. Therefore, the number of parallel kernels (i.e., the kernel pipelines) that can be fitted into the FPGA is increased resulting in higher throughput. Figure 10.4 shows the corresponding kernel speedup – throughput of the approximate kernel normalized to throughput of the exact kernel. As an example, for the n-body kernel the number of kernel pipelines that can be mapped into FPGA is increased from 3 for the exact kernel to 10 for the approximate kernel. Although the maximum clock frequency for this kernel is changed from 209 MHz for the exact kernel to 187 MHz for the approximate kernel, the approximate n-body kernel reaches to 2.98× higher throughput compared to the exact kernel. As another example, convolution kernel reaches 1.41× higher throughput. For this kernel, the number of kernel pipelines (and the maximum clock frequency) is increased from 15 kernel pipelines (and 191 MHz) for the exact kernel to 21 kernel pipelines (and 193 MHz) for the approximate kernel. A geometric mean of 1.82× higher throughput is achieved across

[4]It should be noted that the accelerated profiling process on GPU takes order of milliseconds to determine if the kernel can meet the quality-of-result target. While it takes on average more than an hour to synthesize the approximate OpenCL kernels on Stratix V FPGA.

**Table 10.1** Area utilization for exact (Ext) and approximate (Apx) kernels on StratixV FPGA

| Resource utilization | r-gaussian | | sobel | | n-body | | dct | | conv | | Avg. area reduction (%) |
|---|---|---|---|---|---|---|---|---|---|---|---|
| | Ext | Apx | Ext | Apx | Ext | Apx | Ext | Apx | Ext | Apx | |
| ALUTs | 70668 | 60008 | 72937 | 52213 | 119044 | 58333 | 50530 | 48271 | 53092 | 48573 | 21.5 |
| Registers | 110668 | 90825 | 114892 | 91218 | 172894 | 95156 | 85707 | 80202 | 91062 | 80933 | 20.2 |
| Logic blocks | 27% | 24% | 29% | 23% | 43% | 25% | 22% | 21% | 23% | 21% | 17.3 |
| DSP blocks | 16.4% | 9% | 19% | 19% | 30% | 21% | 25% | 22% | 1.95% | 1.17% | 25.4 |
| Memory bits | 7.91% | 7.09% | 3% | 3% | 8.1% | 4% | 9.56% | 4.39% | 3.7% | 3.36% | 24.8 |
| M20K blocks | 20.23% | 18.4% | 18% | 18% | 25% | 17% | 22.81% | 17.22% | 15.03% | 14.06% | 14.4 |

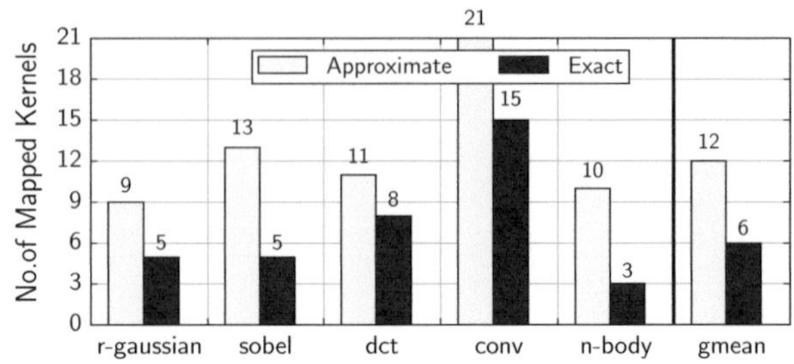


**Fig. 10.3** Number of mapped kernel pipelines on FPGA

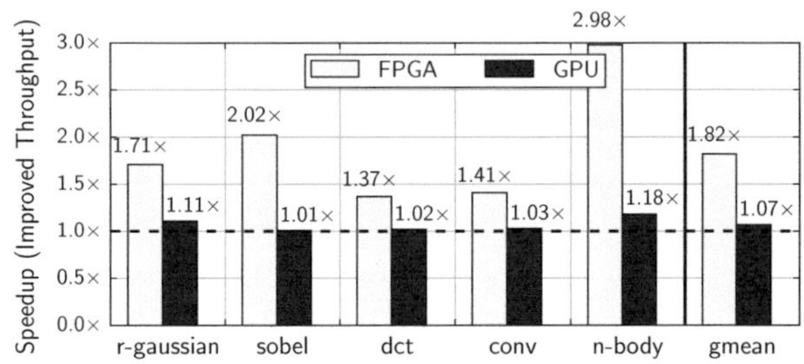


**Fig. 10.4** Speedup with GRATER on FPGA and GPU

these evaluated benchmarks. GRATER increases the number of kernel pipelines until one of the resources reaches its maximum limit. The limiting factor for each of the kernels are as follow: r-gaussian (Logic block, 96%), sobel (DSP block, 91.4%), dct (RAM, 99%), convolution (Logic, 92%), n-body (DSP, 93%). Figure 10.4 also summarizes the speedup for executing the approximate kernels on the GPU, normalized to the exact kernel execution time. The GPU exhibits a maximum speedup of 18% (with a geometric mean of 7%) due to its inflexible pipeline that cannot be fully customized to leverage the precision tuning for boosting throughput per unit of area.

To evaluate the quality loss, we use PSNR for image processing applications and average relative error for the other application domains [13, 14]. We compute the quality loss of each approximate kernel by comparing against the output elements from the exact kernel. For simplicity, here we report the quality loss of all applications by average relative error metric. We set the quality loss target to a maximum of 0.7% for image processing applications (which is equivalent to PSNR of a minimum 30 dB) and 1% for other applications which is conservatively aligned with other work on quality trade-offs [1, 3–5]. We verify the output quality of the optimized approximate kernel with 100 different test input patterns, other than the training input set. Figure 10.5 shows the minimum, maximum, and average quality loss for all the evaluated applications. In all applications, the maximum quality loss is below the required threshold. Hence, it satisfies the target quality-of-result.

The execution time of our proposed algorithm is within few seconds (for sobel and r-gaussian that find the solution without running the genetic algorithm) to few minutes for others.

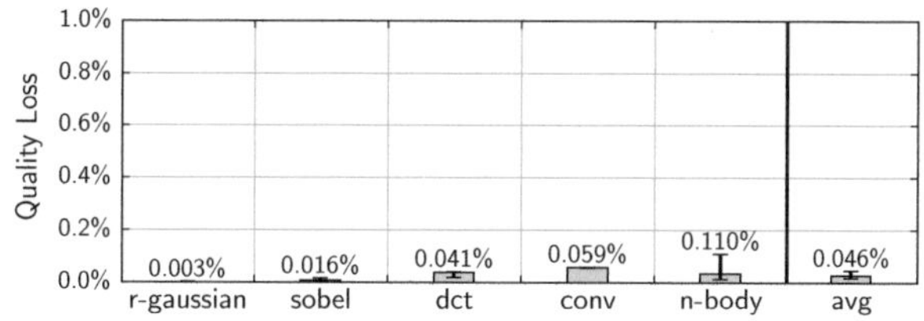


**Fig. 10.5** Quality loss with GRATER

## 10.5 Chapter Summary

This work aims to address the following challenge: *how to exploit approximation in order to increase the benefits of FPGA accelerators without changing the FPGA hardware structure?* Our approach is providing more opportunities for parallel execution by reducing the precision of kernel's data and operations. To this end, we devise GRATER that systematically transforms an OpenCL kernel to an approximate version through a genetic algorithm by reducing its area on the FPGA using the state-of-the-art high-level synthesis tools. The reduction in area results in better utilization of data-level parallelism and thereby increased throughput. The results show that GRATER integrates a larger number of *parallel* kernels on the same FPGA fabric that leads to $1.4\times$–$3.0\times$ higher computational throughput on a modern Altera FPGA with less than 1% loss of quality. GRATER provides these significant benefits without applying any modifications to the underlying FPGA hardware. This feature confirms the efficiency of our framework in exploiting approximation with current hardware platforms. FPGA accelerators provide significant gains in performance and efficiency, yet still require relatively long design cycles to achieve those gains. Automated workflows, such as ours, that improve the benefits of FPGA acceleration are imperative to their widespread applicability.

## References

1. A. Yazdanbakhsh, J. Park, H. Sharma, P. Lotfi-Kamran, H. Esmaeilzadeh, Neural acceleration for GPU throughput processors, in *Proceedings of the 48th International Symposium on Microarchitecture, MICRO-48* (ACM, New York, NY, USA, 2015), pp. 482–493
2. A. Yazdanbakhsh, D. Mahajan, B. Thwaites, J. Park, A. Nagendrakumar, S. Sethuraman, K. Ramkrishnan, N. Ravindran, R. Jariwala, A. Rahimi, H. Esmaeilzadeh, K. Bazargan, Axilog: language support for approximate hardware design, in *2015 Design, Automation Test in Europe Conference Exhibition (DATE)* (2015), pp. 812–817
3. T. Moreau, M. Wyse, J. Nelson, A. Sampson, H. Esmaeilzadeh, L. Ceze, M. Oskin, SNNAP: approximate computing on programmable SoCs via neural acceleration, in *2015 IEEE 21st International Symposium on High Performance Computer Architecture (HPCA)* (2015), pp. 603–614
4. A.B. Kahng, S. Kang, Accuracy-configurable adder for approximate arithmetic designs, in *2012 49th ACM/EDAC/IEEE Design Automation Conference (DAC)* (2012), pp. 820–825

5. P. Kulkarni, P. Gupta, M. Ercegovac, Trading accuracy for power with an underdesigned multiplier architecture, in *2011 24th International Conference on VLSI Design (VLSI Design)* (2011), pp. 346–351
6. Altera SDK for OpenCL. http://www.altera.com/products/software/opencl/opencl-index.html
7. SDAccel. http://www.xilinx.com/products/design-tools/sdx/sdaccel.html (2015)
8. AMD APP SDK v2.9. http://developer.amd.com/tools-and-sdks/opencl-zone/amd-accelerated-parallel-processing-app-sdk/
9. D. Chen, D. Singh, Invited paper: using OpenCL to evaluate the efficiency of CPUS, GPUS and FPGAS for information filtering, in *2012 22nd International Conference on Field Programmable Logic and Applications (FPL)* (2012), pp. 5–12
10. E. Schkufza, R. Sharma, A. Aiken, Stochastic optimization of floating-point programs with tunable precision, in *Proceedings of the 35th ACM SIGPLAN Conference on Programming Language Design and Implementation, PLDI'14* (ACM, New York, NY, USA, 2014), pp. 53–64
11. A.E. Eiben, J.E. Smith, *Introduction to Evolutionary Computing*, 2nd edn., Natural Computing Series (Springer, Heidelberg, 2007)
12. GRATER transcompiler. https://bitbucket.org/act-lab/grater/src
13. S. Misailovic, M. Carbin, S. Achour, Z. Qi, M.C. Rinard, Chisel: reliability- and accuracy-aware optimization of approximate computational kernels, in *Proceedings of the 2014 ACM International Conference on Object Oriented Programming Systems Languages & Applications, OOPSLA'14* (ACM, New York, NY, USA, 2014), pp. 309–328
14. P. Roy, R. Ray, C. Wang, W.F. Wong, ASAC: automatic sensitivity analysis for approximate computing, in *Proceedings of the 2014 SIGPLAN/SIGBED Conference on Languages, Compilers and Tools for Embedded Systems, LCTES'14* (ACM, New York, NY, USA, 2014), pp. 95–104

# Chapter 11
# Memristive-Based Associative Memory for Approximate Computational Reuse

**Abstract** Multimedia applications running on thousands of deep and wide pipelines working concurrently in GPUs have been an important target for power minimization both at the architectural and algorithmic levels. At the hardware level, energy efficiency techniques that employ voltage overscaling face a barrier so-called "path walls": reducing operating voltage beyond a certain point generates massive number of timing errors that are impractical to tolerate. We propose an architectural innovation, called $A^2M^2$ module (approximate associative memristive memory) that exhibits few tolerable timing errors suitable for GPU applications under voltage overscaling. $A^2M^2$ is integrated with every floating point unit (FPU), and performs partial functionality of the associated FPU by pre-storing high frequency patterns for computational reuse that avoids overhead due to re-execution. Voltage overscaled $A^2M^2$ is designed to match an input search pattern with any of the stored patterns within a Hamming distance range of 0–2. This matching behavior under voltage overscaling leads to a *controllable* approximate computing for multimedia applications. Our experimental results for the AMD Southern Islands GPU show that four image processing kernels tolerate the mismatches during pattern matching resulting in a PSNR $\geq$ 30 dB. The $A^2M^2$ module with 8-row enables 28% voltage overscaling in 45 nm technology resulting in 32% average energy saving for the kernels, while delivering an acceptable quality of service. This chapter provides a method for accepting errors in GPUs.

## 11.1 Introduction

There is an ever-increasing demand for multimedia information processing. A graphical processing unit or GPU provides a programmable fabric that orchestrates over 2,000 stream cores to meet the required performance demanded by multimedia applications. Given a limited thermal envelope, powering up over 4 billion transistors makes energy efficiency a primary concern for GPUs. Earlier work has pointed to supply voltage overscaling (VOS) [1, 2] and computational reuse [3] as promising approaches to reduce energy consumption. For a core, there is a voltage and clock frequency operating point at which the core is efficiently functional, but reducing



A. Rahimi et al., *From Variability Tolerance to Approximate Computing in Parallel Integrated Architectures and Accelerators*,
DOI 10.1007/978-3-319-53768-9_11

the operating voltage beyond a critical point leads to so-called "path walls" [4, 5]. The path walls effect is highly pronounced in well-optimized circuits [4]. Hitting the path walls results either in a complete core failure, or massive number of *timing errors* that are very expensive to correct, and wipe out the energy benefits of VOS.

Multimedia applications provide ability to exploit the varying degrees of tolerance to error that an application has due to its programming or inherent application needs [6]. To use this flexibility, "approximate programs," programs that produce results that may be an approximation to the specified results, have an application-dependent fidelity metric to characterize the quality of the output result. For instance, peak signal-to-noise ratio (PSNR) of greater than 30 dB is generally considered acceptable from users perspective in image processing applications. Therefore if program execution is not 100% numerically correct due to few errors during computations, the program can still "appear" to execute correctly. However, recent experiment on an ARM Cortex-M0 core shows that VOS after the critical operating point increases the number of timing errors dramatically [7]. In a similar vein, SRAM-based cache counterpart displays useless behavior under VOS: operating at the nominal voltage is error-free; reducing the voltage down by ~25% generates few errors in data array; below that point there is a massive number of errors in every row and column [8]. This massive number of errors is beyond the capability of the approximate applications to tolerate. Efforts have been done to enable VOS in traditional CMOS-based synthesis by generating approximate hardware blocks for coarse-grained meta-function [9].

In contrast, nonvolatile memories such as resistive RAM (ReRAM/memristor) offers low energy operation with 270 mV–1.0 V [10]. Their downside is limited durability beyond billion write operations that limits their lifetime [11]. Li et al. [12] demonstrate a 1-Mb ternary content addressable memory (TCAM) test chip using 2-transistor/2-resistive-phase-change-storage (2T-2R) cell that achieves >10× smaller cell size than SRAM-based TCAMs, and ensures reliable low voltage search operation. To build energy-efficient GPUs using the CMOS-compatible memristor parts, we have earlier shown integration of the TCAMs with the floating point units (FPUs) for exact computation reuse in Chap. 8. These FPUs consume higher energy per instruction than their integer counterparts, and the overall arithmetic operations contribute to more than 70% of the total GPU power consumption in compute-intensive kernels [13].

Parallel execution in the GPU architectures provides an important ability to combine computational reuse and approximation for reducing energy. This work exploits this opportunity to make three main contributions:

1. We propose approximate associative memristive memory ($A^2M^2$) microarchitectural design to enable simultaneous VOS and computational reuse. $A^2M^2$ is a programmable module accessible by software to store computations that appear frequently, and is tightly integrated to every FPU in the GPU. $A^2M^2$ is composed of a TCAM and a crossbar-based memristor memory block that together represent the pre-stored computations as partial functionality of the associated FPU. Under VOS, $A^2M^2$ exhibits a *controllable* error behavior: when we reduce the voltage

from 1.0 V down to 725 mV, $A^2M^2$ still matches an input search pattern with any of the stored computations within a Hamming distance of 0, 1, or 2.
2. We present a framework, compatible with OpenCL as an industry-standard programming for heterogeneous computing, to profile GPU kernels to identify frequent redundant computations. It applies a fine-grained value partitioning for every FP operation, and extracts a set of values that are occurred frequently through searching the space of possible inputs provided by training samples. The framework carefully pre-stores these key computations in appropriate $A^2M^2$ modules for reusing them to avoid re-executions.
3. We demonstrate the effectiveness of our approach on the Southern Islands GPUs with four image processing kernels adopted from AMD APP SDK v2.5 [14]. We use 10% of Caltech 101 computer vision dataset [15] for the training, and the full dataset for the testing. Our experimental results show that the image processing kernels for all the test images: (1) tolerate the Hamming distance mismatches during pattern matching by displaying a PSNR $\geq$ 30 dB; (2) save on average 32% energy on $A^2M^2$ modules of size 8 made possible by approximate reuse under 28% VOS.

The rest of this chapter is organized as follows. Section 11.2 describes design of $A^2M^2$ for energy-efficient GPU architectures. A framework and kernel execution flow to support $A^2M^2$ is presented in Sect. 11.3. In Sect. 11.4, we explain our methodology and present experimental results followed by conclusions in Sect. 11.5.

## 11.2 GPU Architecture Using $A^2M^2$ Module

### *11.2.1 Southern Islands Architecture*

We focus on one of the most recent GPUs from the AMD, the Southern Islands family (Radeon HD 7000-series). The Southern Islands is based on AMD's Graphics Core Next which is a RISC single instruction, multiple data (SIMD) architecture; it replaces the elder VLIW SIMD architecture from the Evergreen. We target Radeon HD 7970 device which has 32 compute units. Every compute unit contains a scheduler and a set of four SIMD execution units, aka vector units. Each SIMD execution unit has 16 stream cores, or parallel lanes, constituting a total number of 64 stream cores per compute unit.

An OpenCL application is formed of a host program and one or more device kernels that can be run on a GPU device. An instance of the OpenCL kernel is called a work item. Each stream core is devoted to the execution of one work item using the integer or FP units. Most arithmetic operations on a GPU are performed by vector instructions. A vector instruction is fetched once and executed in a SIMD fashion by all its comprising work items. After the fetch and decode stages, the source operands for each instruction are read from vector registers or local memory. The core stage of a GPU is the execute stage, where arithmetic instructions are carried out in each stream core. When the source operands are ready in the vector unit, the execution stage starts to issue the operations into the integer units or FPUs. The execution stage of every FPU has a latency of six cycles and a throughput of one instruction per cycle [16]. Finally, the result of the computation is written back to the destination operands.

## 11.2.2 Approximate Associative Memristive Memory Module

In order to fully exploit the energy saving potentials of both partial memory-based computing and approximate computing, in this section we propose an approximate associative memristive memory ($A^2M^2$) which is tightly integrated to each FPU. The proposed $A^2M^2$ microarchitecture demonstrates controllable approximate computing capabilities under VOS.

For each type of FPU, we first identify the sets of frequent input operands and store them along with their corresponding pre-calculated outputs in an $A^2M^2$ module. Section 11.3.1 describes this flow in details. During the execution, in case of a match between the input values of the FPU and the input patterns stored in $A^2M^2$, the pre-stored results are provided by $A^2M^2$, and FPU re-execution is avoided for frequent operands. $A^2M^2$ module performs the match operation and returns the output at extremely lower energy costs compared to the FPU, thanks to the ultralow power characteristics of memristive memories. This energy cost is further reduced by VOS that relaxes the matching criterion, from the exact to approximate, described in the following.

$A^2M^2$ module consists of two pipelined stages as shown in Fig. 11.1: (I) a memristive TCAM which stores and searches for the high frequent sets of input operands, and (II) a 1T-1R memristive memory which maintains the pre-calculated FPU output results for each set of such frequent operands. For each operation, in the first stage, the TCAM searches to determine whether there is a match between the input operands and the stored operand patterns. In case of a match, the result of the operation is read in the second stage from the corresponding line in the 1T-1R memory.

Each TCAM row stores one set of highly frequent input operands. We use a 2T-2R cell structure for the TCAM design [12]. In this structure each bit of data is stored in a cell that consists of two memristive elements to store the pattern and two access transistors that decouple the memristors from a corresponding match line (ML), as shown in Fig. 11.1. To program the TCAM, the write voltages are applied on the match lines (ML), and access transistors of select devices are connected via the search line (SL) to perform the write operation.

A memristive TCAM operation is based on the fact that a low-resistance path to the ground discharges a precharged line faster than a high-resistance path. Each row in the TCAM has a match line which is precharged during a precharge phase: SLs are deactivated to disconnect the access transistors. During the evaluation phase, one of the access transistors in each bit-cell is ON and connects the ML to the ground via a high- (or low-) resistance path if the pattern-under-search matches (mismatches)

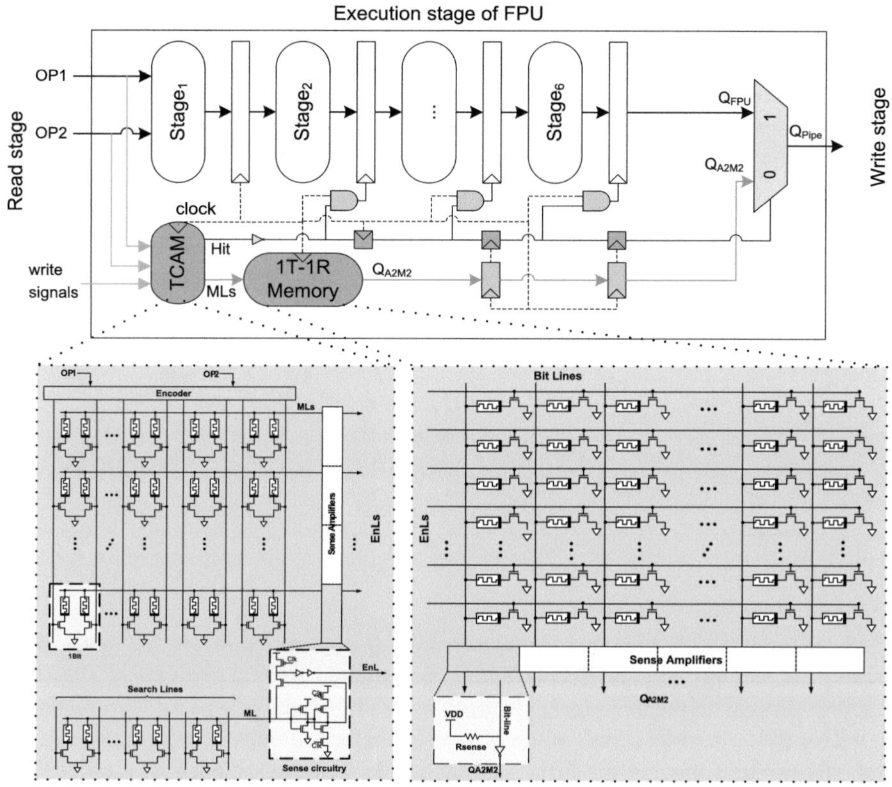


**Fig. 11.1** Execution stage of FPU with $A^2M^2$ module

the stored pattern. In case of an exact match, i.e. bit by bit, the ML stays charged for an extended period of time due to the high-resistance of the memristive device that connects the ML to the ground. If the pattern-under-search and the stored pattern mismatch by even a single bit, the ML will be discharged quickly because of the existence of low-resistive path(s) between the ML and ground, providing a clear margin between an exact match and mismatches. As the number of bit-mismatches increases, the ML will be discharged even faster. A clocked self-referenced sensing circuitry and a 2-bit data encoding scheme is applied [12] to further increase the noise margin and provide a digital match/mismatch output signal. Figure 11.2 illustrates the evolution of the digital "match" signal during the evaluation phase for different number of bit-mismatches based on SPICE simulation results. As it is expected, this signal drops faster when a larger number of bit-mismatches exist. The digital match signals are sampled (i.e., latched) at the end of the evaluation phase. A logic '1' means that the line is not discharged yet, indicating a match. The latched match signals are then fed to the 1T-1R memory stage as enable lines (EnL), to read the previously computed results that are stored in the 1T-1R memory. The logical OR

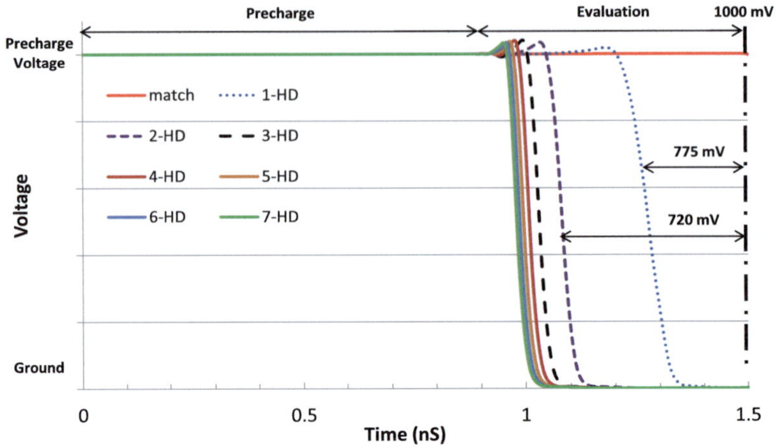


**Fig. 11.2** TCAM match operation under VOS

of the EnLs represents a "hit signal" which indicates that the result is provided by $A^2M^2$ module.

In case of a match, the pre-computed result ($Q_{A^2M^2}$) is read from the memristive memory at negligible energy cost and is propagated toward the end of the FPU pipeline along with the hit signal. The propagated hit signal is used as a clock-gating signal for the remaining stages of the FPU to avoid the redundant computation. Given that only the first stage of the FPU is concurrently working with TCAM, other FPU stages are clock-gated in case of a match which results in considerable amount of energy saving. In case of a TCAM miss, the FPU works normally, and its result ($Q_{FPU}$) is selected as the pipeline output. The hit signal selects whether the $Q_{FPU}$ or $Q_{A^2M^2}$ should be reported as the output.

Figure 11.1 shows the structure of such 1T-1R memory that is used to store the output patterns. To program the memory, a write voltage is applied on the bit-lines, while the enable lines are used to select the target cell. For read operation, the enable lines are derived by the EnL values of TCAM. Assuming an exact matching, either none or only one of EnLs are active at any given clock cycle, connecting the bit-line to the ground through a high-/low-resistance memristive cell, depending on the stored data. The read circuitry works as a voltage divider and is consisted of a sense resistor $R_{Sense}$ and a NOT gate. If the memristor is in the high-resistance state, which represents logic '0', $R_{Memristor} >> R_{Sense}$ and thus the voltage drop on $R_{Sense}$ is negligible and the output of the NOT gate will be a logic '0'. In case of a low-resistance memristor, $R_{Sense} >> R_{Memristor}$, thus most of the voltage is dropped on $R_{Sense}$ and the output of the read circuitry is a logic '1'.

It can be observed in Fig. 11.2 that for few bit-mismatches (e.g. 1 or 2), the drop time of the match signals differ with clear margins. Hence, by shortening the evaluation period (i.e. faster sampling), or similarly reducing the supply voltage while preserving the same evaluation period, a "controllable" approximate matching can be realized in which a pattern with a Hamming distance of 1 or 2 (i.e., the number of bit-mismatches) is considered as a "match". Operating at the nominal voltage of 1 V guarantees an exact matching with 0 number of bit-mismatch. If we reduce the voltage to 775 mV, TCAM also matches the input pattern with any of the stored patterns if there is a Hamming distance of 1 between them (1-HD approximate matching). VOS down to 720 mV matches the input patterns with 2 bit-mismatches (2-HD approximate matching). Further lowering the supply voltages results in an abrupt increase in the number of bit-mismatches.

However, the approximate matching has two downsides: (I) possibility of a false match, and reporting a wrong output as the result of the computation, and (II) having several matches, which would enable several word lines in the 1T-1R memory, resulting in the logical OR of the corresponding outputs being reported as the output of $A^2M^2$ module ($Q_{A^2M^2}$). Possibility of several matches can be avoided if the stored patterns in the TCAM have a minimum Hamming distance (e.g. 3 for 1-HD approximate matching respectively); this is practical given the typical TCAM word size (i.e., 32, 64, or 96), and the small number of TCAM rows. As for the case of a false match, its likelihood is reduced by a proper sizing of $A^2M^2$ module described in Sect. 11.4.2. We limit the match set such that it decreases the likelihood of a false matching and of the introduced error at the same time. In Sect. 11.4.2, we show the application of this approximate matching for different image processing kernels that can tolerate the introduced errors and display a high PSNR while benefiting from the lower energy consumptions. Moreover, $A^2M^2$ module could be designed in a *hybrid* fashion to always exclude the error in a few critical bits (e.g., the sign and exponent bits); for instance, by applying a high voltage to those bits to perform a robust and exact matching, lowering the significance effect of such error.

## 11.3 Framework to Support $A^2M^2$

In this section, we briefly describe our approach to programming $A^2M^2$ and evaluation of $A^2M^2$ effectiveness in improving energy efficiency of GPUs.

### *11.3.1 Execution Flow*

Execution flow using $A^2M^2$ has two main stages: (I) design time profiling, and (II) runtime computational reuse. Figure 11.3 illustrates this execution flow. The goal of profiling stage is to identify redundant computations with a high frequency of occurrence. In the profiling stage, we have an OpenCL kernel, a host code with a training input dataset. We focus on the individual FPUs to observe the dispersion of the input operands at the finest granularity. To expose highly frequent set of operands for each FP operation, we individually profile every type of FP operation and keep the distinct sets of the input operands with the related output result. The output of

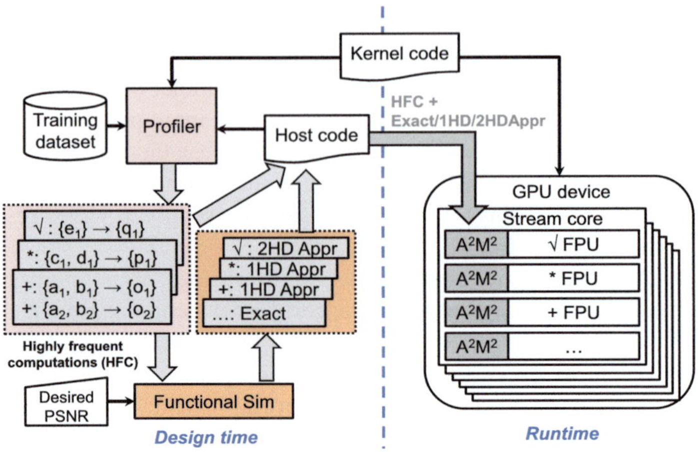


**Fig. 11.3** Execution flow using $A^2M^2$: design time profiling + runtime reuse

this stage for every FP operation is highly frequent computations (HFC): a sorted list of sets of values, each set has the input operand(s) and the related result, and the sets are sorted based on their frequency of occurrence. After extracting HFC, we need to determine how much approximation can be tolerated during the reuse of these key computations. To do so, we leverage the Southern Islands functional simulator to apply different matching constraints for determining the degree of approximation applicable to each $A^2M^2$ module. The simulator starts with the exact matching and then increases the degree of approximation step-by-step by applying 1-HD and 2-HD approximate matching. For every step, the output image is compared with a *golden* output image to measure PSNR. Finally, the maximum degree of approximation is determined for each $A^2M^2$ module such that the introduced errors result in a PSNR higher than the desired PSNR (e.g., 30 dB). This profiling stage is a *one-off* activity whose cost is amortized across all future usage of the kernel.

In the next step, the framework transfers the output of the profiling stage to $A^2M^2$ modules for runtime reuse. The AMD compute abstraction layer (CAL) provides a runtime device driver library that supports code generation, kernel loading, and allows the host program to interact with the stream cores at the lowest level. $A^2M^2$ module are designed to be addressable by software therefore the host code can program them using CAL. Right before launching the kernel execution, the host code programs $A^2M^2$ modules: for every type of FP operation activated during the kernel, a subset of HFC (up to few hundred bytes depending up on the size of $A^2M^2$) in conjunction with the degree of applicable approximation is set for the corresponding $A^2M^2$ modules accordingly. In this way, the framework concurrently programs all the $A^2M^2$ modules integrated to a type of FPU across all the available compute units in the GPU, since their content is equivalent.

### 11.3.2 Design Space for $A^2M^2$

Here, we explain the design space for utilizing $A^2M^2$ modules as a case study for Roberts filter, one of our edge detection kernels. We evaluate the trade-off between the size of $A^2M^2$ module, i.e., the number of rows that store different patterns, with its hit rate. A higher hit rate means higher number of operands are matched with the stored computations in $A^2M^2$ module, therefore there is no need for re-executing the results for those values, leading to higher energy saving. We quantify the hit rate of $A^2M^2$ module for multiply–accumulator (MAC) FPU for 100 test input images. Figure 11.4 summarizes the minimum, the maximum, and the average (shown in bars) hit rates of $A^2M^2$ module with a wider range of sizes. The experiment is repeated for the three matching constraints.

Figure 11.4a shows the hit rates for the exact matching. $A^2M^2$ module with 4-row displays the hit rates in the range of 25–83%. Increasing the size of $A^2M^2$ from 4-row to 8-row, and to 16-row improves the average hit rate from 40 to 42%, and to 50%. Overall, the average hit rates increases less than 12% when the number of rows is increased from 16 to 512. A similar trend of the hit rates versus $A^2M^2$ sizes is observed for the approximate matching, as shown in Fig. 11.4b, c. Once the number of rows is increased from 16 to 512, the average hit rates improves less than 19 and 18% for 1-HD and 2-HD approximate matching, respectively. Figure 11.4 also illustrates that an $A^2M^2$ with a fixed size experiences higher hit rates by switching from the exact matching to any of the approximate matching. For instance, the hit rate of $A^2M^2$ with 4-row increases 12% on average (from 40 to 52%) by using 2-HD approximate matching instead of the exact matching. This increased hit rate is because $A^2M^2$ relaxes the matching constraint therefore more number of input patterns are approximately matched with one of the stored patterns.

In a nutshell, choosing large $A^2M^2$ size has two disadvantage. (I) It diminishes the gain of energy saving, because after a certain size the average hit rates almost saturates, while the energy consumption of the $A^2M^2$ increases for larger sizes. For example, increasing $A^2M^2$ size from 8-row by 64× only brings 25% higher hit rates with 2-HD approximate matching. This significantly lowers the hit rate per unit of power consumed by $A^2M^2$. In Sect. 11.4.2, we show that enlarging $A^2M^2$ beyond a certain size will not bring any energy saving. (II) It increases the likelihood of false matches that might quickly drop PSNR below the desired threshold. Our profiling results indicate that Roberts filter is able to tolerate the errors in computations (an average PSNR of 34 dB) with $A^2M^2$ modules of maximum 512-row using 2-HD approximate matching. Increasing $A^2M^2$ size after 512-row drops the PSNR below 30 dB. Visual depiction and the corresponding PSNR of different matchings for one of the test images are shown in Fig. 11.5.

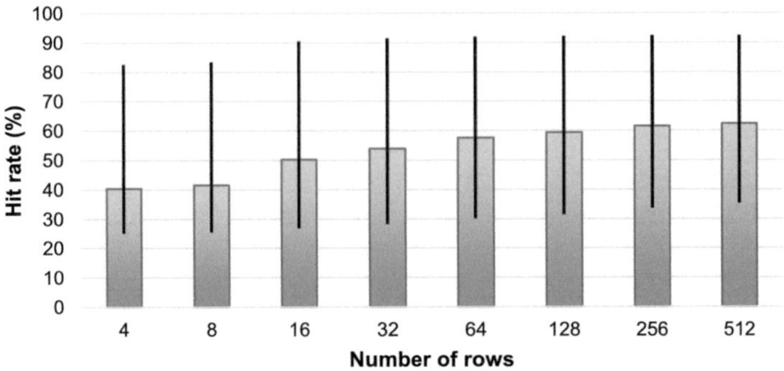


**(a)** Exact matching

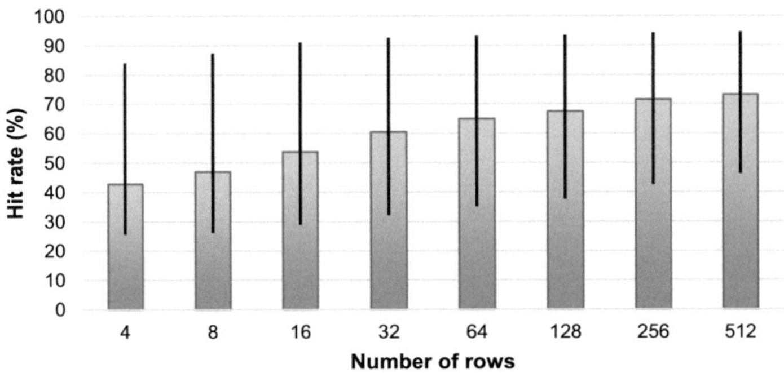


**(b)** 1-HD approximate matching

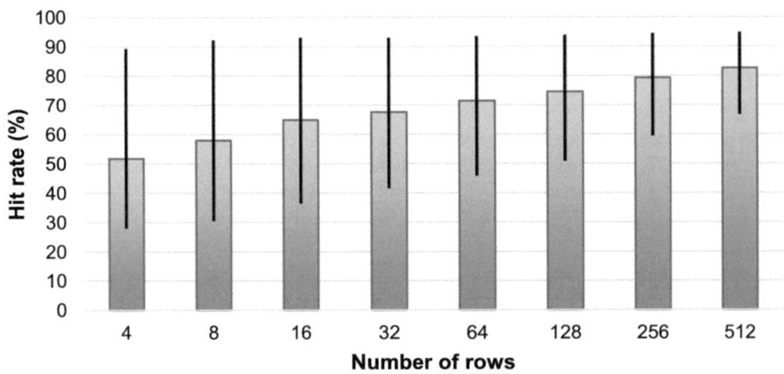


**(c)** 2-HD approximate matching

**Fig. 11.4** Hit rate versus size of $A^2M^2$ for MAC during Roberts filter executions

**Fig. 11.5** Visual depiction of the output quality degradation with exact, 1-HD, 2-HD approximate matching for Roberts filter

## 11.4 Experimental Results

### *11.4.1 Experimental Setup*

We focus on the AMD Southern Islands GPU, Radeon HD 7970 device, but our method can be applied to other GPUs as well. We have adopted image processing applications from AMD APP SDK v2.5 [14] a software ecosystem suitable for stream applications written in OpenCL. We have examined four image processing filters: Roberts, Sobel, Sharpen, and Shift. Multi2Sim [16], a cycle-accurate CPU-GPU simulation framework, is used for profiling and simulations. These kernels typically apply a 2D convolution; we extract frequently activated FPUs during the kernel executions: adder (ADD), multiply (MUL), multiply–accumulator (MAC), and SQRT. Accordingly, the 6-stage balanced FPUs are generated and optimized using FloPoCo [17]. These FPUs are synthesized and mapped using a 45-nm ASIC flow. The front-end flow has been performed using *Synopsys Design Compiler*, while *Synopsys IC Compiler* has been used for the back-end. The FPUs have been optimized for power and a signoff clock period of 1.5 ns. Finally, *Synopsys PrimeTime* is used to report power at the nominal operating voltage of 1.0 V. The second column of Table 11.1 shows the energy per operation for each FPU.

Considering the single precision FPUs, we design $A^2M^2$ module with different word sizes based on the type of FPU. TCAM has a word size of 32-bit for SQRT, 64-bit for ADD, MUL, and 96-bit for MAC; while the crossbar-based memory has a fixed word size of 32-bit for any FPU to maintain the outputs. To estimate power and delay of $A^2M^2$ module, transistor-level SPICE simulations are done using *Cadence Virtuoso*. For the memristor parts, we integrate 50K $R_{on}$ and 50M $R_{off}$ models based on the measurements of fabricated memristors [18]. For the line resistances and capacitances, we use the same model and numbers reported in [19]. Energy operation of $A^2M^2$ modules is shown in Table 11.1. Given the clock period of 1.5ns, $A^2M^2$ modules can reliably work under the designated VOS points (see Sect. 11.2.2). FPUs face massive errors, in this range of VOS, which is simply too high to be useful. We integrate a functional model of $A^2M^2$ module into Multi2Sim that computes the Hamming distance for every FP operation to quantify the hit rates and PSNR drops.

**Table 11.1** Energy consumption (fJ) per operation in 45 nm technology for FPUs and $A^2M^2$

| Module | FPU | $A^2M^2$: exact matching | | | | $A^2M^2$: 1-HD approximate matching | | | | $A^2M^2$: 2-HD approximate matching | | | |
|---|---|---|---|---|---|---|---|---|---|---|---|---|---|
| | | 4-row | 8-row | 16-row | 32-row | 4-row | 8-row | 16-row | 32-row | 4-row | 8-row | 16-row | 32-row |
| ADD | 4742 | 1176 | 1403 | 1858 | 2740 | 644 | 732 | 906 | 1262 | 505 | 555 | 709 | 999 |
| MUL | 9891 | 1176 | 1403 | 1858 | 2740 | 644 | 732 | 906 | 1262 | 505 | 555 | 709 | 999 |
| SQRT | 9983 | 934 | 1137 | 1528 | 2322 | 514 | 594 | 756 | 1084 | 397 | 441 | 593 | 864 |
| MAC | 12051 | 1410 | 1653 | 2122 | 3096 | 774 | 867 | 1052 | 1422 | 612 | 667 | 832 | 1124 |

### 11.4.2 Energy Saving with Corresponding PSNR

Table 11.1 summarizes the energy consumption per operation for individual FPUs, and different sizes of $A^2M^2$ modules in the cases of exact matching, 1-HD, and 2-HD approximate matching. The energy numbers show the potential of $A^2M^2$ modules to reduce the energy consumption per operation. For example for SQRT operation, an exact-matcher $A^2M^2$ module with 8 rows provides $\approx 8\times$ higher energy efficiency compared to FPU counterpart. Although both $A^2M^2$ (exact) and FPU work at the nominal voltage of 1.0V, this energy saving is accomplished through the ultralow power memristive-based computing. The energy saving is further improved by allowing the approximate matching, which improves the energy efficiency by factors of $16\times$ and $22\times$, for 1-HD and 2-HD approximate matching respectively. Such saving trend is consistent for different types of FPUs, and different sizes of $A^2M^2$ modules.

Table 11.1 also demonstrates that increasing the size of the $A^2M^2$ beyond a limit sacrifices the energy efficiency. For instance in case of ADD operation, an exact-matcher $A^2M^2$ module with 64-row roughly consumes as much energy as FPU itself. Any larger $A^2M^2$ module can incur energy penalty rather than improving the energy consumption; since the aggregate energy of integrating FPU with $A^2M^2$ module cannot be paid off by the power saving offered by even an ideal hit rate. In the following, we present energy saving of the kernels using $A^2M^2$ modules with different sizes.

For the four image processing kernels, our framework uses 10% of Caltech 101 computer vision dataset [15] for the training to extract HFC as explained in Sect. 11.3.1. Depending on the size of $A^2M^2$ modules, the framework loads 4, 8, 16, 32, and 64 top pairs of HFC to $A^2M^2$ modules before the kernel execution. We quantify the average energy saving and the corresponding average PSNR degradation over the full dataset [15] as the test cases. Figure 11.6 shows the normalized energy compared to FPUs for each kernel. For all the kernels, the exact-matcher $A^2M^2$ modules with 64-row exhibit poor energy efficiency, for instance Sobel (or Sharpen) faces 20% (17%) higher energy consumption compared to using the normal FPUs. $A^2M^2$ modules with sizes smaller than 64-row provide a significant range of energy saving (16–45%) depending on the size and the degree of approximation. As shown in Fig. 11.6b, $A^2M^2$ modules with 4-row reduce the average energy of Sobel by 20% using 1-HD approximate matching. Increasing the size to 8-row leads to a higher average energy saving of 28% because of the higher hit rate. However, increasing the size beyond 8-row is not optimum because the amount of energy saving offered by the extra hit events is less than the energy overhead due to the increased $A^2M^2$ sizes. We should note that once we reduce the voltage of FPUs down to 775 mV, they face massive number of errors making them impractical to use for low-power computations.

Sobel and Shift kernels cannot tolerate the errors using 2-HD approximate matching, as opposed to Sharpen and Roberts filters. For all the kernels, PSNR is degraded with larger $A^2M^2$ sizes. Increasing the number of stored patterns beyond 32 (or 8) for Sobel (or Shift) abruptly increases the likelihood of a false match that introduces

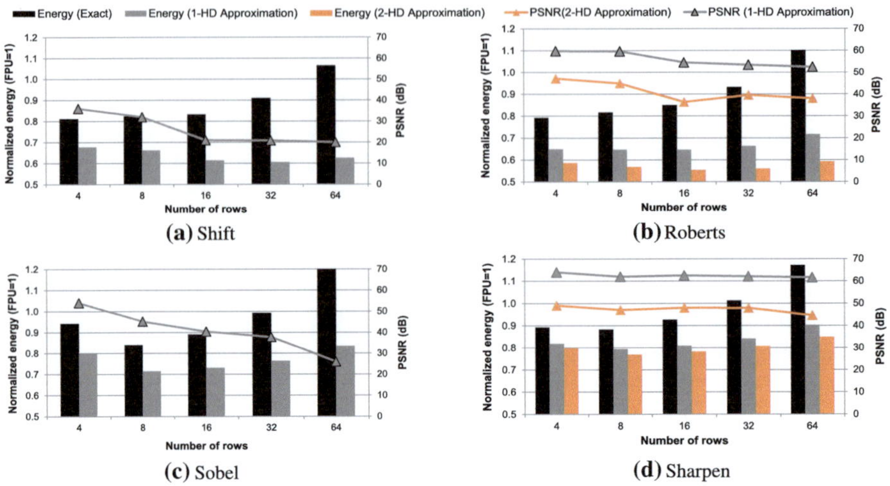


**Fig. 11.6** $A^2M^2$ normalized energy and PSNR: for different sizes, matching criteria, and kernels—values are averaged over the full dataset [15]

more computational errors resulting in a dropped PSNR of 30 dB or lower. Considering the acceptable PSNR of 30 dB or higher, $A^2M^2$ modules with 8-row provide the best average energy saving for Sobel (28%), Sharpen (23%), and Shift (34%); Robert exhibits the best energy saving of 45% with $A^2M^2$ modules of size 16-row. Choosing 8-row as the size of $A^2M^2$ modules brings an average energy saving of 32% across all four kernels, while guaranteeing the acceptable PSNR.

## 11.5 Chapter Summary

We propose $A^2M^2$ as an associative memory module with approximate search capabilities that mixes emerging memristor technology benefits with the application needs to deliver higher energy efficiency. $A^2M^2$ modules are tightly integrated to every FPU to save energy by: (i) recalling the frequent computations therefore avoiding re-executions, and (ii) operating at VOS by accepting the approximate matches. Using the memristor parts in designing $A^2M^2$ enables 28% VOS while incurring up to 2 bits mismatch during the operand matching. We observe that this introduced error into the computations is tolerable by the image processing kernels delivering an acceptable PSNR. Experimental results on the Southern Islands GPU show the integrated $A^2M^2$ modules with 8-row reduce the average kernel energy by 32%. Our continuing work will explore methods to integrate $A^2M^2$ in a programming environment that enables accuracy- and reliability-aware optimizations of approximate kernels.

# References


1. D. Jeon, M. Seok, Z. Zhang, D. Blaauw, D. Sylvester, Design methodology for voltage-overscaled ultra-low-power systems. IEEE Trans. Circuits Syst. II Express Briefs **59**(12), 952–956 (2012)
2. K. He, A. Gerstlauer, M. Orshansky, Circuit-level timing-error acceptance for design of energy-efficient DCT/IDCT-based systems. IEEE Trans. Circuits Syst. Video Technol. **23**(6), 961–974 (2013)
3. W. Wang, A. Raghunathan, N.K. Jha, Profiling driven computation reuse: an embedded software synthesis technique for energy and performance optimization, in *17th International Conference on VLSI Design, 2004. Proceedings* (2004), pp. 267–272
4. S.G. Ramasubramanian, S. Venkataramani, A. Parandhaman, A. Raghunathan, Relax-and-retime: a methodology for energy-efficient recovery based design, in *2013 50th ACM/EDAC/IEEE Design Automation Conference (DAC)* (2013), pp. 1–6
5. J. Patel, CMOS process variations: A critical operation point hypothesis. Online Presentation (2008)
6. M.A. Breuer, Multi-media applications and imprecise computation, in *Proceedings of the 8th Euromicro Conference on Digital System Design, DSD'05* (IEEE Computer Society, Washington, DC, USA, 2005), pp. 2–7
7. L. Lai, P. Gupta, A case study of logic delay fault behaviors on general-purpose embedded processor under voltage overscaling. Technical report, Department of Electrical Engineering, University of California Los Angeles, Los Angeles, CA 90095, August 2014
8. M. Gottscho, A. BanaiyanMofrad, N. Dutt, A. Nicolau, P. Gupta, Power/capacity scaling: energy savings with simple fault-tolerant caches, in *Proceedings of the The 51st Annual Design Automation Conference on Design Automation Conference, DAC'14* (ACM, New York, NY, USA, 2014), pp. 100:1–100:6
9. D. Mohapatra, V.K. Chippa, A. Raghunathan, K. Roy, Design of voltage-scalable meta-functions for approximate computing, in *Design, Automation Test in Europe Conference Exhibition (DATE)* (2011), pp. 1–6
10. M.-F. Chang, J.-J. Wu, T.-F. Chien, Y.-C. Liu, T.-C. Yang, W.-C. Shen, Y.-C. King, C.-J. Lin, K.-F. Lin, Y.-D. Chih, S. Natarajan, J. Chang, 19.4 embedded 1Mb ReRAM in 28nm CMOS with 0.27-to-1V read using swing-sample-and-couple sense amplifier and self-boost-write-termination scheme, in *2014 IEEE International Solid-State Circuits Conference Digest of Technical Papers (ISSCC)* (2014), pp. 332–333
11. H.Y. Lee, Y.S. Chen, P.S. Chen, T.Y. Wu, F. Chen, C.C. Wang, P.J. Tzeng, M.J. Tsai, C. Lien, Low-power and nanosecond switching in robust hafnium oxide resistive memory with a thin Ti cap. IEEE Electron Device Lett. **31**(1), 44–46 (2010)
12. J. Li, R.K. Montoye, M. Ishii, L. Chang, 1 Mb 0.41$\mu$m$^2$ 2T-2R cell nonvolatile TCAM with two-bit encoding and clocked self-referenced sensing. IEEE J. Solid-State Circuits **49**(4), 896–907 (2014)
13. H. Zhang, M. Putic, J. Lach, Low power GPGPU computation with imprecise hardware, in *Proceedings of the The 51st Annual Design Automation Conference on Design Automation Conference, DAC'14* (ACM, New York, NY, USA, 2014), pp. 99:1–99:6
14. AMD APP SDK v2.5. http://www.amd.com/stream
15. Caltech 101 dataset. http://www.vision.caltech.edu/Image_Datasets/Caltech101/
16. Multi2Sim: A heterogeneous system simulator. https://www.multi2sim.org/
17. FloPoCo: Floating-point cores generator. http://flopoco.gforge.inria.fr/
18. K.-H. Kim, S. Gaba, D. Wheeler, J.M. Cruz-Albrecht, T. Hussain, N. Srinivasa, L. Wei, A functional hybrid memristor crossbar-array/CMOS system for data storage and neuromorphic applications. Nano Lett. **12**(1), 389–395 (2012). PMID: 22141918
19. A. Ghofrani, M.A. Lastras-Montano, K.-T. Cheng, Towards data reliable crossbar-based memristive memories, in *2013 IEEE International Test Conference (ITC)* (2013), pp. 1–10,

# Chapter 12
# Spatial and Temporal Memoization

**Abstract** In this chapter, we further combine the methods in detecting and correcting errors with the methods in accepting errors to devise a new hybrid methods Rahimi et al. (IEEE Trans. Circuits Syst. II Express Briefs 60:847–851, 2013) [1], Rahimi et al. (Design, Automation and Test in Europe Conference and Exhibition (DATE), 2014, pp. 1–6, 2014) [2], Rahimi et al. (Temporal memoization for energy-efficient timing error recovery in GPGPU architectures. Technical Report CS2014-1006, Department of Computer Science and Engineering, University of California San Diego, La Jolla, CA 92093, 2014) [3], Rahimi et al. (IEEE Des. Test 33:85–92, 2016) [4]. The cost and speed of error recovery can be improved by *memoization*-based optimization method as a form of computational reuse. Accordingly, we propose two techniques, spatial memoization and temporal memoization, that exploit parallelism in suitable computing fabrics such as GP-GPUs. These memoization techniques exploit value locality and similarity inside data-parallel programs for use in floating-point units (FPUs). Spatial memoization alleviates cost of timing errors recovery, building upon lock-step execution of single-instruction, multiple-data (SIMD) architectures. To support spatial memoization at the level of instruction, we propose a single strong lane, multiple weak lanes (SSMW) architecture. Spatial memoization recalls result of error-free execution of an instruction on the SS lane, and concurrently reuses it to spatially correct any errant instructions across MW lanes. This error correction can be done exactly or approximately. Temporal memoization recalls the context of error-free execution of an instruction on a FPU. To enable scalable and independent error recovery, a single-cycle lookup table (LUT) is tightly coupled to every FPU to maintain few contexts of recent error-free executions. The LUT reuses these memorized contexts to exactly, or approximately, correct errant FP instructions based on application needs. The proposed memoization techniques eliminate the cost of error recovery (e.g., on average 62% for the voltage droop-affected timing errors) and enhance energy efficiency. Spatio-temporal memoization techniques are implemented in standard CMOS technology as a joint method for detecting and correcting with accepting the timing errors in GP-GPUs.

 
A. Rahimi et al., *From Variability Tolerance to Approximate Computing in Parallel Integrated Architectures and Accelerators*,
DOI 10.1007/978-3-319-53768-9_12

## 12.1 Introduction

We have shown in earlier chapters (Chaps. 4, 7, and 9) how a shared memory cluster of processors can schedule parallel work-units to efficiently handle the errors utilizing the fact that runtime system has the ability of "choosing a favor core" in close proximity. On the contrary, such a choice of unit is not available in the data-level parallel architectures where the workload is uniform (SIMD) and all the computing units are fully utilized. Since such architecture has no choice for any alternative execution, it can utilize memoization, or computational reuse, that return a pre-stored error-free result without triggering the error recovery.

Sodani and Sohi [5] introduced the concept of instruction reuse that comes from the observation that many instructions can be skipped if another instance has already been executed using the same input values. The instruction reuse memoizes the outcome of an instruction on hardware tables so a processor can reuse it temporally if the processor performs the same instruction with the same input values. We further extend such notion of temporal memoization to spatial memoization for use in GP-GPUs. GP-GPUs execute workload in SIMD fashion with high utilization. Parallel execution in such SIMD architectures provides an important ability to reuse computation (i.e., memoization) and reduce the cost of recovery from timing errors. We rely on the memoization to safely store the result of a portion of computing on a reliable medium, and then reuse the result rather than re-execution. To do so, we define two notions of memoization at the instruction level: concurrent instruction reuse (CIR), and temporal instruction reuse (TIR). Figure 12.1 shows that for a SIMD architecture:

- CIR answers whether an instruction can be reused spatially across various parallel lanes.
- TIR answers whether an instruction can be reused temporally for a lane itself.

CIR/TIR recalls the result of an error-free execution on an instance of data, then reuses this memoized context in case of meeting a matching constraint. Since different programs exhibit varying degrees of error tolerance, we consider two matching constraints that further extend the application of the memoization to approximate computing domain:

1. Exact matching constraint that enforces full bit-by-bit matching of the single-precision instructions.

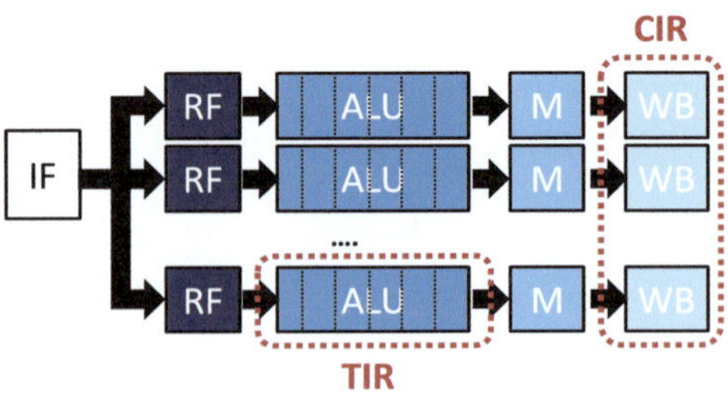


**Fig. 12.1** Concurrent and temporal instruction reuse (CIR and TIR) for SIMD

2. Approximate matching that relaxes the criteria of the exact matching during the comparison by ignoring mismatches in the less significant $N$ bits of the fraction parts.

The latter constraint enables an *approximate error correction* technique suitable for applications in approximate computing to receive further benefits from the memoization technique. In a nutshell, the spatial and temporal memoization techniques leverage inherent value locality and similarity of applications by memoizing the result of an error-free execution on an instance of data; and by reusing this memoized result to exactly (or, approximately) correct any errant execution on other instances of the same (or, similar) data at a very low cost.

These two techniques are fully compatible with the standard CMOS process. In [6, 7], we extend usage of such spatial and temporal reuse techniques in designing associative memory modules (AMMs) by leveraging the emerging CMOS-friendly memristor technology. More details can be found in Chaps. 8 and 11.

## 12.2 Spatial Memoization (Concurrent Instruction Reuse)

To exploit the inherent spatial value locality across SIMD lanes, we propose a SIMD architecture consisting of a single strong lane and multiple weak lanes (SSMW). The SSMW is designed to maintain the lock-step integrity in the face of timing error. The key idea, for satisfying both resiliency and lock-step execution goals, is to always guarantee error-free execution of a strong lane (SS). Then, the rest of weak lanes (MW) can reuse the output of SS lane in the case of timing errors. In other words, SSMW provides an architectural support to leverage CIR for correcting the timing errors of MW lanes.

To measure the exposed spatial value locality over the parallel lanes, we have defined concurrent instruction reuse (CIR) as a metric for the entire kernel execution. CIR is defined as the number of simultaneous instructions executed on the lane1 ($L_1$) through $L_{15}$ of the CUs which satisfy the matching constraint, divided by the total number of instructions executed in all 16 lanes ($L_0$–$L_{15}$). The matching constraint determines whether there is a value locality between the input operands of the instruction executing on $L_0$ and the input operands of another instruction executing on any of the neighbor lanes, i.e., $L_i$, where $i \in [1, 15]$. Thus, a tight (or, relaxed) matching locality constraint ensures that the instructions of $L_0$ and any of $L_i$ are working on the same (or, adjacent) instance of data, and consequently their outputs are equivalent (or, almost equivalent). This exchangeability allows the instructions of $L_0$ to correct any errant output of instructions executing on $L_i$. In the Radeon HD 5870 with 16-wide SIMD pipeline, the maximum theoretic CIR is 93.75% (15 out of 16).

Figure 12.2 shows the CIR rate and the corresponding PSNR for various input pictures while using different matching constraints. As shown in Fig. 12.2c, applying the exact matching constraint yields, on an average, a CIR rate of 27%. This means

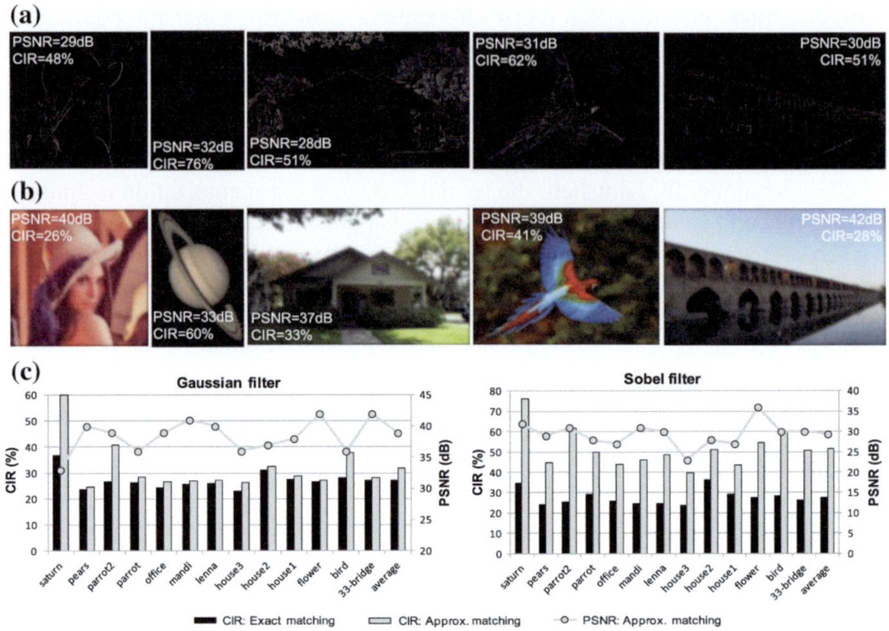


**Fig. 12.2** CIR of the FP with the corresponding PSNR for two kernels. **a** Sobel and **b** Gaussian filters using the approximate matching constraint– 12 bits masked. **c** CIR and PSNR for Sobel and Gaussian filters with the exact and approximate constraints

that 27% of the executed instructions on the whole SIMD can reuse the results of the executed instructions on the $L_0$ (SS lane) for the accurate error correction, without any quality degradation. Approximate matching relaxes the matching criteria through masking the less significant 12 bits of the fraction parts during comparison. Consequently, higher multiple data-parallel values fuse into a single value, resulting in a higher CIR rate for approximate error correction, e.g., up to 76% for Sobel. Applying the approximate matching, on average a CIR rate of 51% (32%) is achieved on the Sobel (Gaussian) filter with the acceptable PSNR of 29 dB (39 dB).

### *12.2.1 Single Strong Multiple Weak (SSMW) Architecture*

The cost of recovery per single timing error on a floating-point SIMD architecture is very expensive. Pawlowski et al. [8, 9] propose to decouple the SIMD lanes through private queues that prevent error events in any single lane from stalling all other lanes, thus enables each lane to recover errors independently. The decoupling queues cause slip between lanes which requires additional architectural mechanisms to ensure correct execution. Therefore, the lanes are required to resynchronize when a microbarrier (e.g., load, store) is reached, therefore, incurs performance penalty.

In response to this deficiency, we exploit the inherent value locality, therefore the SIMD is architected to maintain the lock-step integrity in the face of timing error: SSMW architecture, a resilient SIMD architecture. The key idea, for satisfying both resiliency and the lock-step execution goals, is to always guarantee error-free execution of a lane (SS). Then the rest of lanes (MW) can reuse its output in case of timing errors. In other terms, SSMW provides an architectural support to leverage CIR for correcting the timing errors of MW lanes. Note that to achieve this goal, SSMW *superposes* resilient circuit techniques on top of the baseline SIMD architecture without changing the flow of execution. SSMW employs two circuit resilient techniques. First, it guarantees the error-free execution of the SS lane in the presence of the worst-case PVT variations using voltage overdesign (VO). On the other hand, the MW lanes employ EDS to detect any timing error and propagate an error-bit toward the tail of pipeline stages.

Second, SSMW also employs a CIR detector module for every PE of the MW lanes, as shown in Fig. 12.3. This module checks the matching constraint, and if it is satisfied, the module forwards the output result of the PE in the SS lane to the

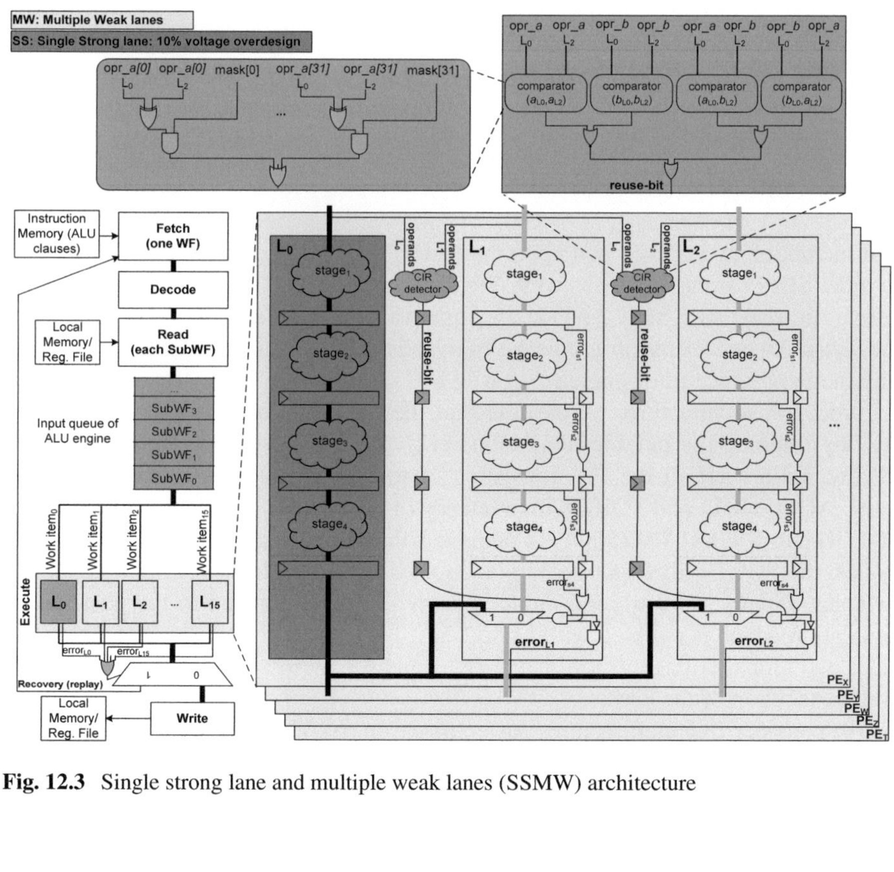


**Fig. 12.3** Single strong lane and multiple weak lanes (SSMW) architecture

output of the corresponding PE in the weak lane. In case of simultaneous matching and timing error for any of the MW lanes, the errant weak lane can reuse the result of SS lane rather than triggering the recovery mechanism. The output result of the SS lane is broadcast via a voltage overdesign network across the MW lanes. The CIR detector module is a programmable combinational logic working on parallel with the first stage of the PE execution; since every PE executes one instruction per cycle, the module is thus shared across all FP functional units of the PE. To check the matching constraint, the module compares bit by bit the two operands of its own PE with the two operands of the PE on the SS lane. All the CIR detector modules share a masking vector to ignore the differences of the operands in the less significant $N$ bits of the fraction part. The masking vector is a memory-mapped 32-bit register that is set by various application demands on the computation accuracy. If the two sets of the operations, with consideration of commutativity, meet the value locality constraint, the module sets a reuse-bit which will traverse alongside the corresponding instruction through the stages of the PE. At the last stage of the execution, the PE takes three actions based on the {reuse-bit, error-bit}. In case of no timing error, i.e., {1/0, 0}, the PE sends out its own computed result to the write stage. If a timing error occurred for the instruction during any of the stages, but it has a value locality with the instruction on the SS lane, i.e., {1, 1}, the PE sends out the computed result of the SS lane, and avoids the propagation of the error-bit to the next stage. Finally, in case of the error and lack of the value locality, i.e., {0, 1}, the PE triggers the recovery mechanism.

### 12.2.2 Experimental Results

Our methodology is developed upon the AMD Evergreen GPUs, but can be applied to other SIMD architectures as well. We use Multi2Sim [10] with naive binaries of kernels in AMD APP SDK 2.5 [11]; the input values for the kernels are generated by the default OpenCL host program. We analyzed the effectiveness of SSMW architecture in the presence of timing errors on TSMC 45-nm ASIC flow. To keep the focus on processor architecture, we assume that the memory components are resilient, e.g., by utilizing the tunable replica bits [12]. We have partially implemented the FP execution stage of the PE, consisting of three frequently exercised functional units: ADD, MUL, and SQRT with a latency of four cycles at the signoff frequency of 1 GHz at (SS/0.81 V/125 °C). To achieve balanced pipelines with latency of four cycles, the SQRT utilizes a polynomial approximation of degree of 5th to decrease its delay. Finally, the variation-induced delay is back annotated to the post-layout simulation which is coupled with Multi2Sim. To quantify the timing error, we consider two global voltage droop scenarios, 3 and 6%, across all 16 lanes during the entire execution of the kernels.

We consider five architectures for comparison. (i) The lane decoupling queues architecture without VO [8, 9]. (ii and iii) SIMD baseline architecture with 10% (or 6%) VO across all 16 lanes. (iv and v) SSMW architecture in which the SS lane, the

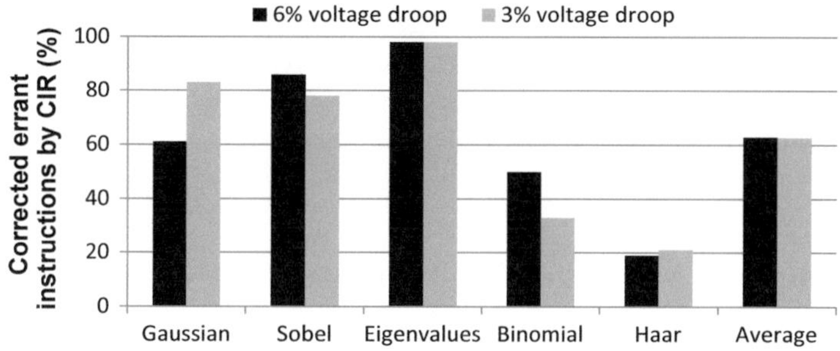


**Fig. 12.4** Effectiveness of CIR for kernels in face of 3 and 6% voltage droops

CIR detector modules, and the broadcast network are guard-banded by 10% (or 6%) VO to guarantee error-free operations. Once SSMW cannot exploit CIR for an error event recovery, it relies on the single-cycle recovery mechanism presented in [8, 9].

To generalize the CIR concept, we have extended our experiments to the error-intolerant applications that do not have inherent algorithmic tolerance. We consider this class of applications as error-intolerant applications that require complete numerical correctness. We have examined three applications where exact matching constraint is applied: Binomial option pricing, Haar wavelet transform, and Eigenvalues of a symmetric matrix. Figure 12.4 shows the effectiveness of SSMW: the percentage of the corrected errant instructions by CIR for all kernels when encountering 6 and 3% voltage droops during the execution. On average for all kernels, SSMW avoids the recovery for 62% of the errant instructions thus significantly reduces the total cost of recovery.

Figure 12.5 shows the total energy comparison of the kernels while experiencing 6% voltage droops. On average, SSMW (10% VO) reduces 8% of the total energy compared to its baseline counterpart. The CIR detector modules increase the delay of the baseline architectures up to 4.9% due to the SS lane broadcast network, and imposes a maximum of 5.7% total power overhead. In comparison with decoupling

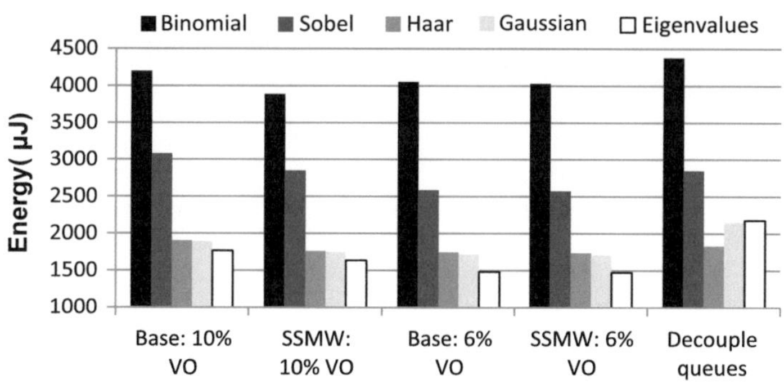


**Fig. 12.5** Energy consumption of kernels in face of 6% voltage droops

queues, SSMW (10% VO) has on average 12% lower energy consumption. The SSMW (6% VO) has also 1% lower energy compared to the baseline with 6% VO, optimistically assuming that the baseline does not incur any timing error while operating at the edge of failure with 6% voltage droops.

## 12.3 Temporal Memoization (Temporal Instruction Reuse)

TIR aims to exploit the value locality and similarity inside each processing element, i.e., FPU in our case. We observe the dispersion of the input operands at the finest granularity for individual FPUs. To expose the value locality for each FPU operations, we consider a private FIFO for every individual FPU. These FIFOs have a small depth and keep the distinct sets of the input operands in the order of instruction arrivals. The FIFO matches a set of incoming input operands and the current content of entries of FIFO using the matching constraint. The FIFO maintains a limited number of recent distinct sets. Therefore, if a set of incoming input operands does not satisfy either matching constraints, the FIFO will be updated by cleaning its last entry and inserting the new incoming operands accordingly.

To exploit the value locality, we tightly couple the FPU pipeline with our proposed temporal memoization module. This module has essentially a single-cycle LUT, and a set of flip-flops and buffers to propagate signals through the pipeline. The LUT is composed of two parts: (i) a FIFO with four entries; (ii) a set of combinational comparators. In every entry, the FIFO maintains a set of input operands and the computed result provided by the output of the FPU in the last stage ($Q_S$). The parallel combinational comparators implement the two matching constraints, and are programmable through a 32-bit memory-mapped register as a masking vector. They concurrently make either a full or partial comparison of the input operands with the stored operands in each entry based on the masking vector. The LUT works in parallel with the first stage of the FPU. Therefore, for every set of input operands, the LUT searches the FIFO to find a match between the input operands and the operand values stored in the entries (i.e., whether the matching constraint is satisfied or not). A match directly results in reuse of results computed earlier. Consequently, this affords the temporal memoization module an opportunity to correct an errant instruction with zero cycle penalty.

### *12.3.1 Temporal Memoization for Error Recovery*

To enable reuse, the LUT propagates a hit signal alongside with the previously computed result ($Q_L$) toward the end of pipeline. The LUT raises the hit signal that squashes the remaining stages of the FPU to avoid the redundant computation by clock gating; the clock-gating signal is forwarded to the rest of stages, cycle by cycle. The stored result is also propagated toward the end of pipeline for the reuse purpose.

**Table 12.1** Timing error handling with temporal instruction reuse

| Hit | Error | Action | $Q_{Pipe}$ |
|---|---|---|---|
| 0 | 0 | Normal execution + LUT update | $Q_S$ |
| 0 | 1 | Triggering baseline recovery (ECU) | $Q_S$ |
| 1 | 0 | LUT output reuse + FPU clock-gating | $Q_L$ |
| 1 | 1 | LUT output reuse + FPU clock-gating + masking error | $Q_L$ |

The hit signal selects the propagated output of the LUT ($Q_L$) as the output of the FPU; it also disables the propagation of timing error signal (if any) to the recovery unit, thus avoids the costly recovery. Therefore, each hit event reduces energy by locally retrieving the result from the LUT, rather than doing full re-execution by the FPU. In case of a LUT miss, the FIFO is updated to maintain the last recently computed values. It is implemented through a write enable signal ($W_{en}$) that ensures there is no timing error during execution of all stages of the FPU for computing $Q_S$. Finally, if simultaneous timing error and miss occurred, the error signal will be propagated to the recovery unit that triggers the baseline recovery. Table 12.1 summarizes these four states.

### *12.3.2 Experimental Results*

We focus on the execution stage consisting of six frequently exercised functional units: ADD, MUL, SQRT, RECIP, MULADD, FP2FIX. We select eight kernels form AMD APP SDK 2.5 [11]. For these applications, TER avoids costly recovery that improves the energy efficiency with an average energy savings of 8% (for 0% timing error rate) to 28% (for 4% timing error rate). The memoization techniques are explained in detail in [1–3].

## 12.4 Chapter Summary

We propose architectures to enable spatial and temporal memoization techniques that seek to reduce error recovery costs by reuse of concurrent and temporal instructions, while maintaining a lock-step execution of the SIMD architecture. These proposed memoization techniques exploit the value locality and similarity in data-parallel applications that are explicitly exposed to the parallel lanes. These memoization techniques recall result of an error-free execution on an instance of data; then reuse the memoized result to exactly (or, approximately) correct any errant execution on other instances of the same (or, similar) data. Together, they significantly reduce the cost of resiliency and enhance the range of variability-induced timing errors that can be mitigated at very low cost. On an average, the proposed SSMW eliminates the cost of recovery for 62% of the voltage droop-affected instructions, and reduces 12% of the total energy compared to recent work [8].

The observations in this chapter open an opportunity to exploit instruction reuse technique, in the context of memristive associative memories, to spontaneously apply clock gating to FPUs beforehand, therefor avoiding redundant computations.

## References

1. A. Rahimi, L. Benini, R.K. Gupta, Spatial memoization: concurrent instruction reuse to correct timing errors in simd architectures. IEEE Trans. Circuits Syst. II Express briefs **60**(12), 847–851 (2013)
2. A. Rahimi, L. Benini, R.K. Gupta, Temporal memoization for energy-efficient timing error recovery in gpgpus, in *Design, Automation and Test in Europe Conference and Exhibition (DATE), 2014* (2014), pp. 1–6
3. A. Rahimi, L. Benini, R.K. Gupta, Temporal memoization for energy-efficient timing error recovery in GPGPU architectures. Technical Report CS2014-1006, Department of Computer Science and Engineering, University of California San Diego, La Jolla, CA 92093 (2014)
4. A. Rahimi, L. Benini, R.K. Gupta, CIRCA-GPUs: increasing instruction reuse through inexact computing in GP-GPUs. IEEE Des. Test **33**(6), 85–92 (2016)
5. A. Sodani, G.S. Sohi, Dynamic instruction reuse, in *Proceedings of the 24th Annual International Symposium on Computer Architecture*, ISCA '97, ACM, New York, NY, USA (1997), pp. 194–205
6. A. Rahimi, A. Ghofrani, M.A. Lastras-Montano, K.-T. Cheng, L. Benini, R.K. Gupta, Energy-efficient GPGPU architectures via collaborative compilation and memristive memory-based computing, in *Proceedings of the The 51st Annual Design Automation Conference on Design Automation Conference*, DAC '14, ACM, New York, NY, USA (2014), pp. 195:1–195:6
7. A. Rahimi, A. Ghofrani, K.-T. Cheng, L. Benini, R.K. Gupta, Approximate associative memristive memory for energy-efficient GPUs, in *Proceedings of the 2015 Design, Automation & Test in Europe Conference & Exhibition*, DATE '15 (2015), pp. 1497–1502
8. R. Pawlowski, E. Krimer, J. Crop, J. Postman, N. Moezzi-Madani, M. Erez, P. Chiang, A 530mv 10-lane SIMD processor with variation resiliency in 45nm SOI, in *2012 IEEE International Solid-State Circuits Conference Digest of Technical Papers (ISSCC)* (2012), pp. 492–494
9. E. Krimer, P. Chiang, M. Erez, Lane decoupling for improving the timing-error resiliency of wide-SIMD architectures, in *Proceedings of the 39th Annual International Symposium on Computer Architecture*, ISCA '12, IEEE Computer Society, Washington, DC, USA, (2012) pp. 237–248
10. Multi2Sim: a heterogeneous system simulator. https://www.multi2sim.org/
11. AMD APP SDK v2.5. http://www.amd.com/stream
12. A. Raychowdhury, B.M. Geuskens, K.A. Bowman, J.W. Tschanz, S.L. Lu, T. Karnik, M.M. Khellah, V.K. De, Tunable replica bits for dynamic variation tolerance in 8T SRAM arrays. IEEE J. Solid-State Circuits **46**(4), 797–805 (2011)

# Chapter 13
# Outlook

**Abstract** Microelectronic variability is a phenomenon at the intersection of microelectronic scaling, semiconductor manufacturing, and how electronic systems are designed and deployed. Using timing errors, as the most threatening manifestation of variability, we showed various levels of microelectronic circuit and system design where the effects of variability can be mitigated. Increasing leakage power is another challenge; variability has already had a major impact on the leakage power. Coordinated combined methods are central to an emerging outlook on variability tolerance as discussed below (Fig. 13.1).

## 13.1 Domain-Specific Resiliency

### *13.1.1 Software*

Software presents a great unexploited potential for diagnosis and mitigation of variation effects. Software requires *runtime* monitoring and *re-calibration* to approach to the edge-of-failure or "nothing works" for energy efficiency, but never go on the other side of the border with failure. The key point is that at *design time* there is not enough knowledge and there is too much variability and sensitivity to have a viable design time approach. A self-learning approach can discover the frontiers of efficient operating points, of course we need a means of recovery is something goes bad. Distributed software techniques and paradigms will therefore become increasingly pervasive even at the chip level. The trend should be toward avoiding global variability bottlenecks, through arranging a mix of redundant execution (avoiding single-point of failure), globally asynchronous communication and orchestration, and fine-grained rollback.

 
A. Rahimi et al., *From Variability Tolerance to Approximate Computing in Parallel Integrated Architectures and Accelerators*,
DOI 10.1007/978-3-319-53768-9_13

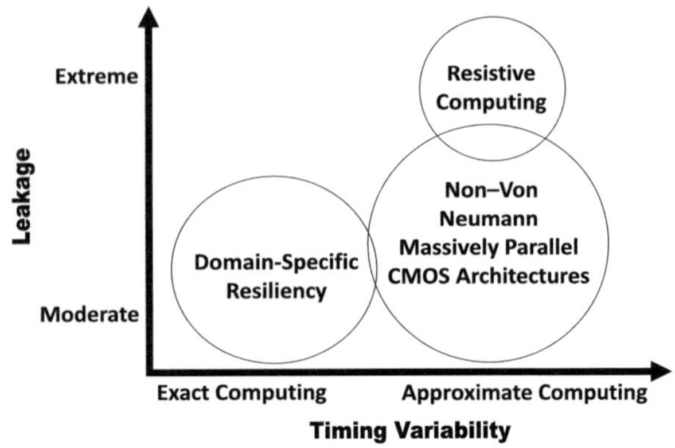


**Fig. 13.1** Emerging outlook on variability tolerance

### *13.1.2 Architecture*

Variability mitigation is about cost and scale. Modular and scalable architectures such as those found in the programmable accelerators enable better observability and controllability of variations through explicit parallelism. Both hardware and software can enhance variability tolerance by tuning two available axes: *configurations* and *choices*. Hardware and software can jointly configure available settings of an architecture and appropriate parameters explicitly coded in applications. They can also selectively choose a suitable hardware resource, or an alternative code path. For instance, one alternative can select an optimized approximate kernel rather than exact one results in significant resource reduction enabling integration larger number of parallel kernels on the fixed budget the underlying architecture.

### *13.1.3 Circuit*

Focusing on CMOS circuits, a large spectrum of asynchronous circuits can be utilized. For a given sub-circuit (either exact or approximate), a synthesis tool would have the choice of selecting a *communication* scheme among available different communication templates for realizing that sub-circuit. In other words, the problem of determining the level of accuracy of a sub-circuit will be transformed to *how much* energy we want to spend on ensuring the sub-circuit functional *integrity* instead of spending the energy on the actual sub-circuit computation.

## 13.2 Non-Von Neumann Massively Parallel Architectures

Emerging applications including graphics, multimedia, web search, data analytics, and cyber-physical system go beyond primarily numerical computations for scientific use to interacting with sensory interfaces. Functional non-determinism presents in these applications at human–cyber interfaces. In this direction, we have observed limitations on Von Neumann architecture: we can only *relax* the execution stage and fine-grained mechanisms incur high overhead with increased complexity. On the other hand, we found out that parallel architectures and parallelism in general provide the best means to combat and exploit variability to design resilient and efficient systems. Therefore, fast, highly scalable, and space-efficient methods are very desirable that could initiate a departure from Von Neumann architecture toward neuro-inspired computing. For instance, new sparse and distributed data representation promises to deliver substantial energy advantages and robust operation. Further, utilizing resistive memory elements not only solves the leakage problem but also provides a dense memory-centric architecture suitable for neuro-inspired resistive computing.

# Index





A. Rahimi et al., *From Variability Tolerance to Approximate Computing in Parallel Integrated Architectures and Accelerators*,
DOI 10.1007/978-3-319-53768-9

MIX
Papier aus verantwortungsvollen Quellen
Paper from responsible sources
FSC® C105338
FSC
www.fsc.org

If you have any concerns about our products,
you can contact us on
**ProductSafety@springernature.com**

In case Publisher is established outside the EU,
the EU authorized representative is:
**Springer Nature Customer Service Center GmbH**
**Europaplatz 3, 69115 Heidelberg, Germany**

Printed by Libri Plureos GmbH
in Hamburg, Germany